D0217927

THE CURIOUS HISTORY OF RELATIVITY

THE CURIOUS HISTORY OF RELATIVITY

How Einstein's Theory of Gravity Was Lost and Found Again

Jean Eisenstaedt

TRANSLATED BY ARTURO SANGALLI

PRINCETON UNIVERSITY PRESS
PRINCETON AND OXFORD

Published by Princeton University Press, 41 William Street,
Princeton, New Jersey 08540
In the United Kingdom: Princeton University Press,
3 Market Place, Woodstock, Oxfordshire OX20 1SY

ISBN-13: 978-0-691-11865-9
ISBN-10: 0-691-11865-5

Library of Congress Control Number: 2006926977

British Library Cataloging-in-Publication Data is available

Ouvrage publié avec le soutien du Centre National du Livre—Ministère Français
Chargé de la Culture

This book has been published with financial aid of the CNL
(Centre National du Livre), French Ministry of Culture

This book has been composed in Palatino

Printed on acid-free paper. ∞

pup.princeton.edu

Printed in the United States of America

1 3 5 7 9 10 8 6 4 2

Contents

Foreword

EINSTEIN IS THE twentieth century's most famous scientist. His face is universally recognized, thanks to a well-known photograph showing the unconventional side of a white-haired scientist with a child's sense of humor. Everyone also knows that his most famous equation, $E = mc^2$, has something to do with nuclear energy and that Einstein was concerned with the social and political consequences of nuclear weapons. But few know that the new concepts introduced by Einstein shattered the foundations of science and created most of the conceptual framework that allowed and guided the development of twentieth-century physics. Even fewer would be able to explain those revolutionary concepts and the way they underlie present-day physics.

This book was written for all those who wish to understand, without undue technical detail or mathematical formalism, the nature of this sweeping conceptual revolution. Under Jean Eisenstaedt's expert but benevolent guidance, motivated readers will be able to follow Einstein's line of thought step by step. They will experience the joy of reliving the critical moments of the twentieth century's most brilliant mind. At a time when scientific popularization usually offers only a faint echo of the intellectual battles behind any nontrivial advancement of scientific thought, readers will be able to witness the extreme difficulties that Einstein and his contemporaries faced when they were led to demolish the longstanding notions of (Euclidean) space and (Newtonian) time and to replace them with the new concept of a curved and dynamic space-time.

Jean Eisenstaedt's beautiful book describes with all the required intellectual subtlety and historical rigor both the global strategy and the detailed "battle plans" of the Einsteinian conquests. The author allows us to relive Einstein's "imaginative art," which was often based on visual or dynamic intuitions that took him years to formalize and transform into precise scientific theories. Contrary to what a quick survey of Einstein's achievements may suggest, his great ideas did not come out of his mind ready-made. They matured slowly, following a long intellectual journey filled with ambiguities, doubts, and errors before the path toward clarity finally appeared.

Readers will have the pleasure to discover Einstein's "style"—his focus on the new "principles" that was common to all his discoveries. Besides an in-depth introduction to Einstein's thought, this book provides a rare insight on the history of the receptivity to Einstein's ideas, its first verifications, and developments. For instance, I believed I knew the essentials of the history of the experiments, inspired by the general theory of relativity, to measure the deflection of light by the Sun. Thanks to this book, I had the pleasure to discover a story much more complex than I had thought, full of twists and turns, with its interplay of human passions and political factors. Readers will be reassured as they discover that scientific research is not (only) an abstract undertaking, but also an enterprise carried out by men and women with their passions and weaknesses. Jean Eisenstaedt does not hide Einstein's own weaknesses and those of his contemporaries. In particular, he analyzes in detail the "prehistory" of the concept of black hole and how most of the great scientists of the first half of the century got bogged down by a new version of Zeno's paradox, remaining stuck in a certain intellectual horizon that prevented them from "inventing" the modern concept of the black hole. Likewise, readers will have the pleasure to discover how cosmology forced scientists to push to the limit the great Einsteinian idea of a curved and dynamic space-time, far from the "neo-Newtonian" coziness in which the relativistic description of the Earth's cosmic backyard so often slumbered.

I hope that these few preliminary remarks, indicative of the merits of this book, will persuade the readers to let Jean Eisenstaedt's expert hand guide them along the paths in Einstein's space-time.

Thibault Damour
Professor at the Institut des Hautes Études Scientifiques
Member of the Academy of Sciences

Acknowledgments

As the English translation of this book goes to press, I would like to thank all those—too many to be named here—who have read, perused, commented on, or corrected earlier versions of the text, discussed a particular idea, or suggested changes at some point or another of its production, especially Thibault Damour, who has also written a very enthusiastic foreword. I could not forget my colleagues at the numerous libraries where I had the pleasure to work: not everything can be found on the Internet!

I also wish to thank the publisher, Princeton University Press, for the quality of their work on this English version, as well as CNRS-Éditions for the original French edition.

Finally, I would like to express my heartfelt and friendly thanks to Arturo Sangalli for the remarkable interest and care he demonstrated all along during the translation of this book, besides his help in correcting a few inaccuracies.

As for the mistakes and the preconceived ideas so difficult to shake off, they are naturally exclusively mine. There is still work to do!

Jean Eisenstaedt

INTRODUCTION

A Difficult Theory

A new scientific theory does not triumph by convincing its opponents and making them see the light, but rather because its opponents die, and a new generation grows up that is familiar with it.

Max Planck[1]

GENERAL RELATIVITY, that is, Einstein's theory of gravitation, has long been considered incomprehensible. There are many reasons for that opinion, and if they are certainly not all technical, they are not merely ideological either. Rethinking space-time—accepting that geometry is not the one our senses (and our education) have taught us and that the universe is curved—requires a true intellectual effort.

Things had gotten off to a bad start with special relativity, which was not an easy theory either, to say the least. In 1959, four years after Einstein's death, the distinguished theoretical physicist Max von Laue revealed to Margot Löwenthal, Albert's daughter-in-law, his difficulty in understanding Einstein's 1905 article on special relativity and the forty years it had taken him to succeed: "[S]lowly but steadily a new world opened before me. I had to spend a great deal of effort on it. . . . And epistemological difficulties in particular gave me much trouble. I believe that only since about 1950 have I mastered them."[2]

This admission of Max von Laue, a Nobel physicist familiar with Einstein's work and author of some excellent books on relativity, should help us, as we begin our journey through *The Curious History of Relativity*, to accept our own difficulty in approaching relativity. We are not alone in this situation. Many before us have faced similar obstacles and have resisted relativity's ideas, logic, and consequences—and made plenty of mistakes which, I hope, will help us to better understand the theory.

The difficulty in understanding "the" theory of relativity (special and general relativity were often confused) was so widespread at the turn of the century that it gave rise to a story, probably apocryphal but soon reaching mythical proportions, in which only three persons could understand Einstein's theory. But it appears that the myth was based on a true story. . . .

On 6 November 1919, at Carlton House in London, the extraordinary meeting of the Royal Society devoted to the results of the English expeditions and chaired by J. J. Thomson has just ended: general relativity has been "verified." Eddington, the hero of the day and the center of attention, chats with his colleagues. Ludwik Silberstein, a small, bearded man, well-known relativist, and author of a decent treatise on special relativity, who also had an inclination for debate and heated discussions and was very sure of himself and his quick mind, joins the group and exchanges a few polite words with an amused Eddington. The atmosphere is light, full of jesting remarks. Silberstein then asks Eddington:

> Isn't it true, my dear Eddington, that only three persons in the world understand relativity?" Silberstein confidently expects the obvious, polite reply, "But, apart from Einstein, who, my dear Silberstein, who, if not you . . . and I, if you allow me."
>
> Eddington, however, remains aloof, silent, amused. Silberstein insists: "Professor Eddington, you must be one of the three persons in the world who understand general relativity." To which Eddington, unruffled, replies, "On the contrary, I am trying to think who the third person is!"[3]

More than two centuries earlier, a student passed Newton on a Cambridge street and observed in a hushed voice: "There goes the man who has written a book that neither he nor anyone else understands."[4]

Definitely, gravitation does not appear to be an accessible subject. And yet, relativity is not as difficult to understand as public rumor has it, nor is it the only theory to resist comprehension or to make us wonder. By learning about the difficulties experienced by the brightest, we may perhaps more easily accept our own and come to terms with the limits of our understanding. In short, we may progress.

NOTES

1. Max Planck, quoted in Feuer, 1982, p. 87.

2. Max von Laue to Margot Einstein, 23 October 1959; quoted in Holton, 1973, p. 198.

3. Quoted in Chandrasekhar, 1979, p. 216.

4. From Christianson, 1984, p. 291.

CHAPTER ONE

The Speed of Light and Classical Physics

KINEMATICS, the science of the relationship between space and time, is the study of the motion of particles without regard to its causes. This science lies at the foundations of physics, to which it gives shape and support.

Until 1905, *c*, the speed of light, did not appear in the equations of kinematics or in those of gravitation, but it was already present in Maxwell's equations. The introduction of *c* into kinematics was a fundamental step, one that marked the transition from classical, or Galilean, kinematics to relativistic kinematics, or special relativity. The next step would introduce *c* into the equations of gravitation, and general relativity would be born.

Light played a central role in these developments. Einstein had always been troubled by the problems posed by light. When he was only sixteen he was already wondering what would happen if he were to "pursue a beam of light with the velocity of *c* (velocity of light in a vacuum)."[1] A difficult question, as naïve questions often are. Would the wave appear unmoving, frozen in the ether? Or would it simply continue to travel, unperturbed? He concluded that "however, there seems to be no such thing, whether on the basis of experience or according to Maxwell's equations."

Let us begin with some classical experiments dating back to Newton's time—specifically, some attempts to measure *c* made at the time of the discovery of the "retardment of light," a way of expressing the fact that the speed of light was finite, as Ole Römer had just shown.

FROM RÖMER TO BRADLEY

The year was 1667. Louis XIV reigned with Colbert as his Minister of Finance. The Observatoire de Paris, headed by Jean Dominique Cassini, had just been created and, in order to improve the quality of the country's maps, it was expected that the excellent astronomical observations made by Tycho Brahe a century earlier at Uraniborg would permit a better determination of longitudes. But before they could be used, Brahe's data had to be converted to the meridian of Paris. It was therefore essential to know the exact difference in longitude between the two observatories. To this end, Jean Picard traveled to Denmark, where he was assisted by Ole Christensen Römer, a twenty-seven-year-old Dane who was already studying Brahe's observations.[2] Favorably impressed by the young Römer, Picard brought him back to Paris in the summer of 1672. Römer was lodged at the Paris Observatory and actively participated in the scientific life of the city. Before long, he was appointed tutor of the dauphin—the heir to the throne—and was granted an allowance by Louis XIV. Beside his astronomical duties, which he conducted with exemplary zeal, Römer took part in various other projects, in particular that of supplying water to Versailles.

Tycho Brahe's observations and those made during Picard's trip were soon available. To proceed with the project, it was necessary to observe a given star at the same time from the two places whose difference in longitude was to be established. The eclipse[3] of one of Jupiter's moons observed simultaneously by the two astronomers would provide the "beep" guaranteeing the simultaneity of the measurements. Tables predicting the times of the eclipses were available, but wide discrepancies had been discovered between the times predicted by Cassini's tables and the actual observations—up to several minutes, which seemed incomprehensible given the precision of the measurements. Cassini's predictions for the eclipses of Jupiter's first moon Io, from August to November 1676 were published in the *Journal des Sçavans* in August of the same year. Römer "announced to the Academy at the beginning of September that, if his hypothesis was correct, an emersion of the first moon which should take place on 16 November would occur 10 minutes later than the time predicted by the usual calculations."[4] On 9 November 1676, the emersion was in fact observed by Picard at 5 hours, 37 minutes, and 49 seconds, a delay of a little over 10 minutes from Cassini's calculation.

Römer then turned his attention to the tables for the eclipses of one of Jupiter's moons. He was a convinced Copernican and therefore believed that satellites must obey Kepler's laws: their eclipses and their recurrent and cyclic immersions and emersions ought to be regular, periodic. The tables should have revealed an absolute periodicity of the eclipses of any given moon, but that was not quite the case. Let us pass over the discussions and controversies of that period, because other hypotheses besides "the retardment of light" would be proposed, in particular, that the irregularity in the apparent motion of the moons could be due to a perturbation caused by another moon. Römer, for his part, believed that the irregularity was related to the retardment of light, to the fact that "light is not transmitted instantly," as they still said at the time. In fact, ignoring the irregularities due to the other moons, the motion of Io is perfectly periodic: eclipsed by Jupiter's cone of shadow, it disappears at very regular intervals 1.769 days apart. But that is not the image we receive, because this image depends on the (variable) distance at the time of the observation. In fact, the distance from Jupiter to the Earth varies widely throughout the year, because it is a function of the motion of both planets around the Sun. If, at a given moment, the moon is closer or farther from the Earth than at some other instant (because of Jupiter having moved toward the Earth or away from it), the image of this event—immersion or emersion—carried by light, will reach us earlier or later. We must therefore distinguish between the moment of the moon's emersion and that of its observation. If the distance between Jupiter and the Earth were always the same, or, equivalently, if the interval of time between the emersion and its observation were constant, we would receive the film of the event after some delay (less than twenty minutes—the time that light takes to travel from Jupiter

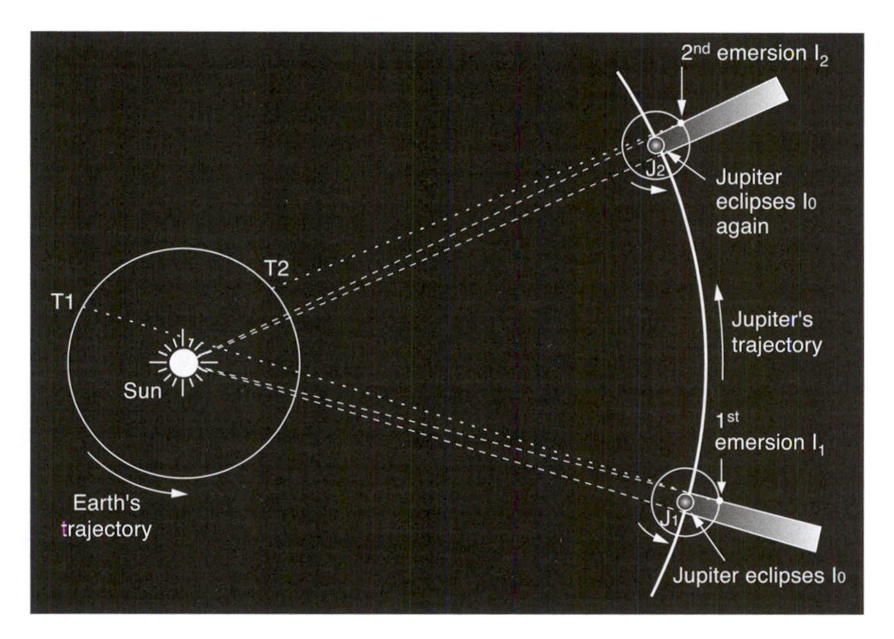

Figure 1.1. In 1676, Römer performed a first measurement of the speed of light at the Paris Observatory.

to us), but this film would not be distorted. But since the distance between Jupiter and the Earth varies, this variation (which may reach several minutes) will be reflected on the *apparent* tables of the observed phenomena.

On 21 November 1676, Römer delivered to the Academy of Sciences his paper in which he "demonstrates that to travel a distance of about 3,000 leagues, which corresponds approximately to the diameter of the Earth, light requires less than one second of time" as follows from "the observations of Jupiter's first moon."[5]

The satellite Io (I) is eclipsed by Jupiter's shadow cone. Two eclipses of Io, I_1 and I_2, are represented six months apart. In the first one ($T_1 - I_1$), the Earth (T) is farther away from Io than in the second ($T_2 - I_2$). The traveling time of light will be different and will affect the tables of eclipses.

Less than one year after its publication in the *Journal des Sçavans*, a translation of Römer's article appeared in the *Philosophical Transactions of the Royal Society*, and the retardment of light found in John Flamsteed, Astronomer Royal at the Greenwich

Observatory near London and to whom Römer would pay a visit in 1679, a convinced advocate. Flamsteed probably communicated with Newton and argued in favor of Römer's case—an easy task, since Newton would have little difficulty accepting an idea that was a necessary consequence of his own theories. But it would take Bradley's discovery of aberration some fifty years later to really convince everybody of the "retardment of light"—that the speed of light is finite.

At the end of the 1720s, James Bradley, professor of astronomy at Oxford and future successor of Edmond Halley (of comet fame) as Astronomer Royal, was trying to observe the phenomenon of parallax, that is, the variation in the apparent position of a fixed star resulting from the change in the position of an observer on Earth. It is a simple problem of perspective: if the Earth really revolved around the Sun, as it was believed after Copernicus, the image of the sky should vary with the seasons—the stars close to the Earth should change their position with respect to those farther away. A phenomenon that, being the touchstone of the Copernican hypothesis, is as necessary as it is difficult to observe.

Bradley failed to observe any parallax, but he discovered another, very curious phenomenon: the annual aberration, which, in some sense, also revealed the annual motion of the Earth: the fixed stars appeared to trace out on the lens a very small ellipse, a twin sister or projection of the Earth's trajectory around the Sun. Bradley quickly came up with an explanation: the speed of the light coming from the fixed stars is combined with that of the Earth.

Since the annual aberration involves the ratio between the seasonal variation of the Earth's speed and the speed of light, any variation in the speed of the light coming from a star should be detectable by a change in the angle of aberration. Now, the most accurate measurements available since the eighteenth century indicated a strong constancy of the annual angle of aberration, which suggested that the speed of the light emitted by the stars and measured by an observer on Earth was constant. But how to reconcile this experimental fact with the (assumed) constancy of speed of the light emitted by the stars, with the (probable) speed of the stars themselves, and with the daily, as well as annual, variations in the speed of the observer on Earth? In short, the obvious variations in the relative speed of the observer with respect to the source should have an impact on the speed of the light measured on Earth, and therefore on the angle of aberration. Something was not right. How was it possible that the speed of light appeared to be constant?

ARAGO'S TRAVEL READINGS

In 1806, François Arago was twenty years old. Fresh from the recently founded École polytechnique, he had already completed an important piece of work, "On bodies' affinity for light," in collaboration with Jean-Baptiste Biot, an accomplished physicist. At the end of that year, they both set out on a mission to Spain. In order to extend Cassini's meridian plane, they carried out the geodesic measurements needed to draw the meridian axis centered at the Paris Observatory, based on which a map of Spain would be produced. Arago's reading material for the trip was neither a novel nor a book of poetry, but Newton's *Opticks*.[6] We may easily imagine him at the summit of some Spanish mount reading and annotating his copy of *Optiks* while he waited for the optimal meteorological conditions to carry out the topographical readings. Like any other physicist, astronomer, or philosopher of his time, he was a convinced Newtonian.

Arago's trip would be an eventful one. He was taken prisoner by the North Africans and returned to France only three years later. As soon as he arrived, in 1809, he got on with his work and began a series of measurements whose purpose was to reveal the difference in speed of the light coming from a given star depending on whether the observer is moving toward it or away from it. Since the speed of the Earth with respect to a star changes direction due to the Earth's daily rotation, it is enough to observe the star in the morning and again in the evening. Suppose that, in the northern hemisphere, we observe, just before sunrise, a star situated west of us; since our telescope is being carried by

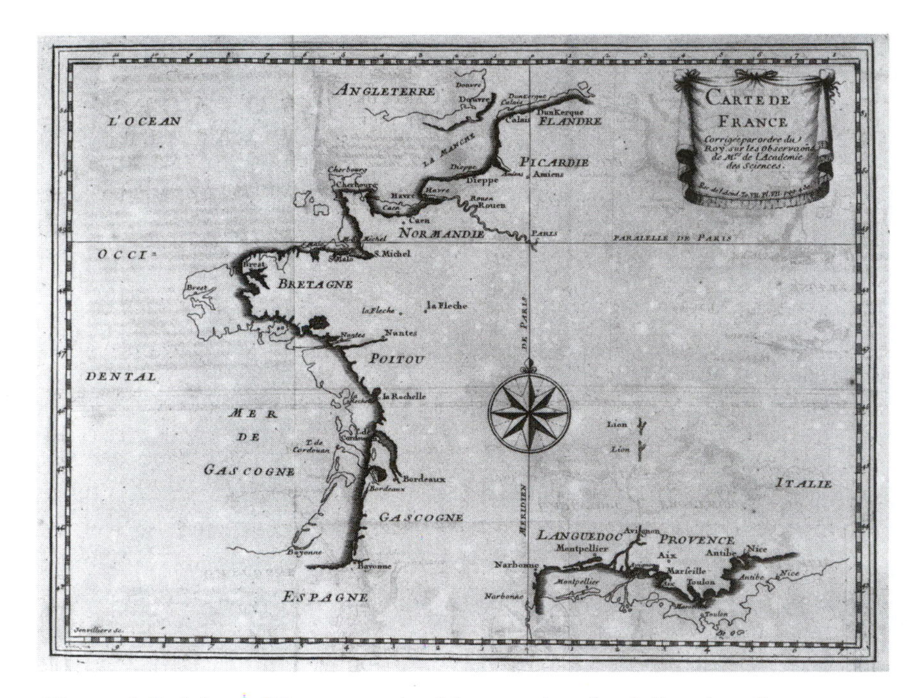

Figure 1.2. Map of France revised by royal order following the Observations made by the Members of the Academy of Sciences. In 1693, under Picard's direction and thanks to the new astronomical methods, the contour of the kingdom was surveyed for the first time. It was then realized that the coast of Brittany was considerably closer to the capital than previously thought, to the great displeasure of Louis XIV, who, seeing his territory become smaller would have declared, not without humor: "These gentlemen of the Academy, with their precious calculations, have taken part of my kingdom away from me". (Photograph courtesy of Observatoire de Paris)

the Earth's rotation toward the east, the speed of the light particles coming from the star should be equal to their emission speed of the light minus our rotation speed. In the evening, with the telescope pointing east, we observe the same star; its light should now have a speed equal to the emission speed plus the tangential speed of our telescope, which is rotating along with the Earth.

But how to measure the speed of light? Arago employed a prism to analyze the light coming from the star. Newton's optics would

Figure 1.3. View of the Observatoire de Paris, southern side. Engraving reproduced from *Observations astronomiques faites à l'Observatoire de Paris*, 1837–1846. It is here that Arago tried to detect the differences in the speed of light. (Photograph Observatoire de Paris)

then provide the theoretical foundation for the interpretation of the results. In Newton's optics—then known as emission theory—the angle of refraction as the light exits the prism was a function of the speed of the incoming light particle. Arago therefore expected the angle of refraction of the beam of light to vary from morning to evening or between two observations made six months apart. But none of this happened; the angle of refraction was exactly the same morning and evening, and hence the speed of light appeared to be the same, regardless of whether the observer was moving toward the source or away from it. It was a bizarre, nonsensical result. How was it possible that the relative speed of a light particle was not the vectorial sum (i.e., obtained using the law of the parallelogram) of the emission speed, the observer's speed, and the speed of the source? Unless . . . Could

Figure 1.4. Portrait of Arago by S. M. Cornu (1840). Astronomer, physicist, and politician, Arago (1789–1853) was one of the figures of the Republic during the Romanticism. (Photograph Observatoire de Paris)

it be that Newton's optics was incorrect and refraction was not a function of the incoming light particle's speed?

Arago hesitated for a long time before questioning Newton's optics, considered untouchable. On the other hand, discarding the law of the addition of velocities in kinematics was unthinkable. And so, it was only later that he finally decided to abandon Newton's emission theory in favor of the work on wave optics by Augustin Fresnel, an engineer in the Department of Bridges and Roads with a passion for optics. Following a series of experiments, notably those of the English physician Thomas Young, many other results connected with the discovery of interference and polarization also pointed in the same direction—to a wave theory of light. A supporter of the monarchy, Fresnel took a stand against Napoleon, who had returned from the island of Elba. During the One Hundred Days—the period between the emperor's return to Paris and his second abdication—Fresnel was exiled to Normandy, where he could pursue his beloved work. And so, slowly, Newton's particle optics would give way to Young's and Fresnel's wave optics.

FRESNEL: A PHYSICS OF LIGHT

To explain Arago's negative result, Fresnel introduced the ether,[7] a subtle fluid filling all space but which only manifested itself through the propagation of light. He assumed that light was a wave transmitted by an ether that was partially dragged by refringent mediums,[8] even if it was undetectable.

To be sure, not all Newtonians immediately admitted defeat, in particular in France. It would take Hippolyte Fizeau's experiment, in the middle of the nineteenth century, to convince the last skeptics. According to Newton's emission theory, the speed of light had to be greater in water than in the air, whereas Fresnel's theory predicted the exact opposite. Fizeau's was then a clear and crucial, if difficult, experiment which, by definitively refuting the optics of Newton, confirmed that of Fresnel. Throughout the nineteenth century, Fresnel's theory would allow physicists

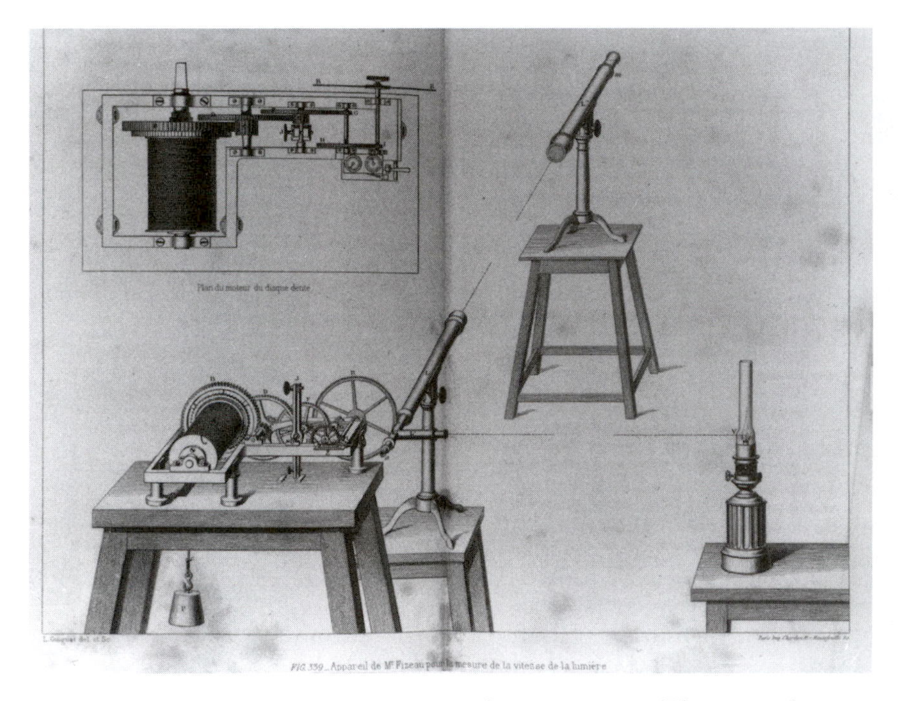

Figure 1.5. Fizeau's experiment according to Arago. (Photograph Observatoire de Paris)

to explain the laws of reflection and double refraction, interferences, the laws of diffraction, polarization, the colored rings phenomenon and, with some difficulty, also aberration.

In the wake of Arago and Fresnel we find Babinet, Stokes, Hoek, Hertz, Fizeau, Foucault, Mascart, Doppler, Angstrom, Maxwell, and Lorentz—an impressive list of names.[9] From then on, wave optics ruled, and hundreds of articles, dozens of experiments, and a huge quantity of work ensued. In the second half of the nineteenth century, Maxwell, and later Lorentz, would make the connection between light and electromagnetism.

Albert Michelson's experiment at the end of the nineteenth century was a sort of replay of Arago's. The purpose was still the same: to compare the speed of two light rays, but, almost a century later, the theoretical and technical context was completely different. Arago's optics had been abandoned in favor of that of Maxwell-Lorentz's, which extended and improved Fresnel's theory, and it was no longer a question of revealing a difference in speed by a difference in the angle of refraction. Seeking to improve the accuracy of the measurements, Michelson used interferometry, a sophisticated technique. The two arms of the interferometer were perpendicular to each other; if one of them was then pointed in the direction of the Earth's motion, the speed of the two light beams would be different, resulting—it was expected—in different traveling times and, hence, interferences. The experiment was performed several times, and the amazing result—that no effect at all could be observed—drove Michelson to despair and plunged the world of physics into a crisis. Far from being partially dragged, the ether was at rest with respect to the Earth. Michelson's result became well known around the world, and it can now be found in every college physics text.

To call into question the law of the addition of velocities and Galilean kinematics was absolutely unthinkable (we can never emphasize this enough). A nineteenth-century man or woman simply could not contemplate such a possibility, one which contradicted their most immediate perceptions, their representation of the physical world. For anyone who had observed the motion of trains—which were then in rapid development—how could it be possible to accept that speeds were not *really* additive, that the speed of light in a vacuum was a constant, regardless of the speed of the observer or that of the source?

Before accepting the reality of the nonadditivity of velocities, one had to become accustomed to electric, magnetic, and light phenomena involving much greater speeds. It would be necessary to get a feeling for higher and higher speeds, to enter into a world where a different physics was emerging—the physics of light and electromagnetism; the physics of electric motors, for example, required one to imagine (and hence also to see) bodies in motion in a magnetic field. The first electric motor was built in 1821 by Michael Faraday, based on H. C. Oersted's discoveries in electromagnetism, and the first electric dynamo appeared ten years later. Heinrich Hertz's experiments on the speed of propagation of electrodynamic actions date from 1888. It would require time, the time for all that to sink in, to simply get used to the speed of electrons. And it would be necessary to be pitilessly confronted by reality in order to accept changing something so important, so strong, so *obvious* (also and foremost from a psychological point of view) as Galilean kinematics—to accept changing the relationship between space and time. The relativistic revolution consisted primarily in this radical modification of the notion of motion, in this loss of a classical physics that was believed to be well understood and that had to be abandoned.

KINEMATICS AS THE SCIENCE OF MOTION

Let us return once again, but from a different angle, to the logic behind the experiments mentioned above. It was natural to expect that the speed of light should depend on the speed of the observer measuring it as well as on that of its source. If we see passing by us that famous train Einstein immortalized, the speeds that we measure from the platform (that of the ball rolling on the train's floor or that of the beam of light from a flashlight held by a dumbfounded passenger) must take into account the train's motion and hence its speed with respect to the platform. We know that, in the simplest case where those speeds have the same direction, they are added together: the speed of the ball with respect to the platform will be the sum of its speed with respect to the train plus that of the train relative to the platform. There is *additivity* of speeds. That is a fact so simple that it appears simplistic. It is based on classical kinematics, the science that describes the apparent motion of particles without reference to its causes, the science of the cinema of particles.

Until the end of the nineteenth century, classical kinematics was Galilean kinematics, which only brought into play space and time and the related concepts of velocity and acceleration. Kinematics is the science of motion—of motion in space and in time; of the exchange of information in space and time—in the universe. It is clearly the first of the physical sciences because the whole of physics makes use of it; it is the science of the foundations of physics.

The physical space in question is what we will call, for convenience, a *Newtonian* space, which from a mathematical point of view, is a Euclidean space of dimension three (essentially, width, length, and height) plus an absolute time, hence the famous four dimensions which, as a matter of fact, existed before Einstein. Three dimensions in a space where an ordinary and absolute distance is defined, together with a time that is itself absolute. "Absolute" simply means "totally independent with respect to any phenomenon." This allows us to define an ordinary speed: distance divided by time; hence the theorem, also commonplace, of the addition of speeds. In such a Newtonian framework, the speed of Einstein's passenger with respect to the platform is that of the passenger with respect to the train (as he—passengers in Einstein's time were always men—walks along the moving train from the rear to the front) plus the speed of the train relative to the platform. And all this can be rigorously proved, provided we assume that the space is really "Newtonian" and that speed is nothing but distance divided by time; in short, provided we accept those (too) simple and obvious definitions.

Let us take the time to go over this proof in order to convince ourselves of its obviousness and hence to better appreciate the rift that its calling into question will create—to better understand how difficult it will be for the physicists to accept that so clear a conclusion was not right.

A passenger boards a one-hundred-meter-long train at the rear. The train travels one hundred kilometers in one hour. Our passenger gets off the train at the front; he has therefore traveled one hundred kilometers and one hundred meters while, during the same interval of time, the train (i.e., the front of the train) has traveled only one hundred kilometers. Thus the speed of our passenger will be one hundred kilometers per hour—the speed of the train—plus one hundred meters per hour (or, equivalently, 0.1 km per hour, his speed relative to the train), that is, 100.1 kilometers per hour. In short, all this amounts to a simple addition: $100.1 = 100 + 0.1$. There is indeed addition of speeds. Q.E.D.

This little argument holds the key to all *classical* kinematics. If, as Michelson's experiments showed, this theorem of the addition

Box 1. The Addition of Velocities

We shall only consider the case where the velocities (or speeds) are parallel.

Let V_1 be the speed of the train with respect to the platform, and V_2 the speed of the ball rolling on the train's floor.

What is the speed V of the ball with respect to the platform?

Elementary, my dear Watson! V is the sum of V_1 and V_2: $V = V_1 + V_2$.

Elementary . . . only in classical kinematics! For it is no longer true in relativistic kinematics; it can be shown that in this case c comes in and

$$V = (V_1 + V_2)/[1 + V_1V_2/c^2].$$

Velocities no longer add.

Of course, only trains and balls are involved here, hence both V_1/c and V_2/c are small compared to 1 (the speed of a high-speed train is four million times smaller than that of light, not to mention the speed of a ball). We can then neglect the term V_1V_2/c^2 and so the classical addition formula is accurate enough. On the other hand, if we are dealing with particles in an accelerator, relativistic kinematics is essential.

Notice that from this formula it follows that c is a limit speed. If our passenger is holding a flashlight, then the speed of the emitted photon is $V_2 = c$, but the speed of that photon with respect to the platform is also c:

$$V = (V_1 + c)/[1 + V_1c/c^2] = c.$$

This is not a paradox. It is simply the fact that velocities do not obey a classical addition law but rather an addition law that is a bit more complex.

of speeds is not valid, in particular for light, then something is not right with our initial assumptions. Is it absolute space? Absolute time? The definition of speed? Where's the mistake? It must necessarily be in the concepts and definitions of physics, for the rest is just mathematics!

The most convincing solution physicists will find will be special relativity. Not much will remain of our initial hypotheses: neither Newton's absolute time nor the definition of speed will survive. But, above all, in this new kinematics a new physical constant will appear, *c*. It will no longer be possible to add two speeds without the intervention of *c*. No kinematics will be possible without *c*; no physics will be possible without *c*.

However, a qualification of the above statements is in order. No physics of fast motions (or of high energy) will be possible without the intervention of *c*, but for phenomena involving low speeds, classical kinematics remains of course valid. The new kinematics will only apply if the speeds involved are close to the speed of light, for *c* comes into play through a correction factor "v/c." This is clearly seen in the new formula for the addition of velocities (box 1).

LORENTZ'S ELEVEN HYPOTHESES

At the end of the nineteenth century, Hendrik Antoon Lorentz, professor at the University of Leyden, in the Netherlands, formulated no less than eleven hypotheses to explain those phenomena—a theory, or theoretical framework, that was called *the electromagnetism of moving bodies*.[10] This was a lot of hypotheses. Henri Poincaré, who was working on the same problems, said of Lorentz: "If he got away with it, it is only by piling up hypotheses."[11] To which Lorentz replied: "Surely this course of inventing special hypotheses for each new experimental result is somewhat artificial. It would be more satisfactory if it were possible to show by means of certain fundamental assumptions and without neglecting terms of one order of magnitude or another, that many electromagnetic actions are entirely independent of the motion of the system."[12]

Lorentz was at the time the best informed specialist on the question, with his knowledge of the data, the theories or theoretical attempts, and the various experiments and their possible explanation. He probably knew too much and was firmly caught

Figure 1.6. From left to right: Albert Einstein, Hendrik Antoon Lorentz, and Arthur Eddington. (© AIP Emilio Segrè Visual Archives)

in the web of classical arguments, in the system of nineteenth-century physics. In this story, he played the role of a conservative, while Einstein was the revolutionary who boldly challenged the fundamental concepts of classical physics. In short, it was a conflict of generations.[13]

Simplified in the extreme, such was the picture that Einstein—a fervent admirer of Lorentz, by the way—faced at the beginning of the century, a picture of which he ignored many facts, and it is just as well. Here we shall do likewise. Sometimes ignorance is a good thing. "Better a well-formed head than a well-filled one," a saying that applies quite well to Einstein, for it took a free spirit to dare redraw the portrait of physics. We will try to follow him!

And so, at the beginning of the twentieth century, fundamental physics was divided into two large sections. On the one hand, sound, light, and electrodynamics fell under the wave approach; on the other, Newtonian mechanics was restricted to material particles with the exclusion of light. Within Newtonian mechanics we find kinematics (the mathematical description of motion), of which the symbol is Galileo, and dynamics (the study of motion in conjunction with the forces that cause it), which is the work of Newton. Newtonian mechanics, that monument of modern science that, not entirely by chance, is called *Principia*,[14] is based on principles named axioms and definitions, not all of them clearly stated but nevertheless obvious to everyone. Finally, Newton's theory of gravitation was part of dynamics. At the beginning of the century, it was a theory that seemed to defy eternity. Even today (see chapter 10) it can more than adequately describe the basic dynamics of the solar system, but it is completely inadequate to deal with astrophysics and cosmology (chapters 14 and 15).

And so we can, a posteriori, understand the difficulties that physicists encountered all through the nineteenth century. They had to build a bizarre, convoluted, *fabricated* perspective, one which was in a sense created from scratch: a physics that would be able to reconcile phenomena involving greater and greater speeds (optics and later electromagnetism) with a kinematics that had run out of steam, a kinematics in which *c* had no place, a kinematics demanding, as we have seen above, an addition of velocities that reality rejected. In midcentury, and in an amazing way, Maxwell had taken the plunge and (unconsciously) abandoned Galileo's kinematics by introducing a theory in which *c* already played a fundamental role, timidly appearing as an important constant in physics.

Then Fresnel's and Lorentz's theories, and beside them a large body of work by nineteenth-century scientists, acted as a buffer between Galilean and Einsteinian kinematics, and it is in that sense that they were unnatural, *fabricated*. But they remained, nevertheless, important theories, for they allowed physicists to make progress in a difficult field and to account for numerous phenomena. In a certain sense, that is the real task of the fundamental physicist, who must provide a solution to problems that are, at least for a while, inextricable. That was indeed the case and we can, a posteriori, appreciate the difficulty of the problem those physicists faced.

Why sometimes everything becomes clear—as would be the case in 1905—is a mystery. But perhaps this enlightenment is the result of a larger point of view that makes things appear simpler than they did before, when a narrower view had to be offset by some artifice. We have seen, for example, the extent to which, long before Fresnel, *c* was already implicitly present. But this relative simplicity has a price: we must give up some obvious facts, such as the addition of velocities and the Euclidean structure of space. It is a price that not everybody was prepared to pay, so that this enlightenment would not be shared by all.

It follows from all this that in order to better *understand*, to more distinctly hear special relativity, we must not listen too closely to Fresnel or Lorentz and must interpret Arago's experiment in a simpler way than its author did. Here we must *forget* history! That is precisely what the physicists did, as they recast and founded anew their view of the physical world—a recasting that required them to forget the old foundations of their discipline and its meanders.

NOTES

1. Einstein, 1949, in Schilpp, 1949, p. 53.

2. On Ole C. Römer's (or Olaus Roemer's) works, see Taton, 1978.

3. The disappearance of the moon behind the planet. The instant this disappearance takes place is called "immersion," while that of reappearance is known as "emersion."

4. B. Le Bouyer de Fontenelle, 1676, quoted by S. Débarbat, in Taton, 1978, p. 146.

5. Römer, 1676. Römer's article, as well as an analysis of his observations, may be found at the end of S. Débarbat's, in Taton, 1978, pp. 151–154. From the estimate of one French common league, the speed of light according to Römer may be derived, a value of about 215,000 kilometers per second, approximately 30 percent less than modern measurements.

6. Newton, 1730.

7. On the history of the physics of light at that time, see Buchwald, 1989.

8. Refringent: literally "refracting." These are media such as water, glasses, or crystals whose index of refraction is different from that of vacuum, which is by convention equal to 1.

9. The most important articles on the beginnings of relativity were published in Lorentz et al., 1923, in particular the fundamental papers by Einstein, Minkowski, and Lorentz.

10. On this topic, see Holton, 1973, in particular pp. 165–183 and 185–195.

11. Quoted in Holton, 1973, p. 187.

12. Lorentz, 1904, quoted in Lorentz et al., 1923, p. 13.

13. As in the title of Feuer's 1974 book.

14. Newton, 1687. English translation, Newton, 1729.

CHAPTER TWO

Light and the Structure of Space-Time

Insofar as mathematical propositions refer to reality, they are not certain, and insofar as they are certain, they do not refer to reality.

A. Einstein[1]

LET US EXAMINE now the reasons that led Einstein to special relativity. His doubts did not stem primarily from experimental considerations, for Lorentz's theory roughly managed, if not to explain, at least to account for most of the phenomena. What he could not bear was the complexity, the ad hoc and contrived character of those explanations. He was probably already convinced that the laws of physics must be simple, which was certainly not the case of the electrodynamics of the time. It is not surprising then that in his famous 1905 article[2] he started from an incoherence, a strange asymmetry between a phenomenon and its description: "We know that Maxwell's electrodynamics—as we usually conceive it today—leads, when applied to bodies in motion, to certain asymmetries that do not appear to be inherent in the phenomena."[3]

It was just the opening of his article, and he already showed his colors! But what was his problem with those asymmetries? In the then classical explanations of that phenomenon there was *indeed* (we can today say in hindsight) something really disturbing. The question was that of describing "the electrodynamical interaction between a magnet and a conductor." What amazed Einstein—as well as other physicists, it must be said—was essentially the fact that the question was not treated in the same way if it was the magnet rather than the electrical conductor that was in motion with respect to the ether. Wasn't that odd? An electric force was claimed to be in action if the magnet was in motion, but a magnetic one was supposed to intervene if it was the conductor that moved. The theoretical interpretation of the phenomenon was not the same in both cases; it was not symmetric. That was strange, a sign that the theory did not fit the facts, that it was not adequate. But this was not the only important point for, in the next paragraph, Einstein called attention to the vain attempts to detect a motion of the Earth with respect to the ether. The impossibility of detecting such a motion: this was a fact that certainly struck him.

Einstein proposed quite simply to change the point of view, that is, to take the results of experiments, real ones or thought experiments, at face value. He was now convinced that he was dealing with a fundamental property and that any experiment designed to observe the motion of the Earth would be unsuccessful. From there to conclude that the concept of absolute rest was an empty one there was but one step, which he took: "To the concept of absolute rest does not correspond any property of phenomena." Exit the ether; out with that ghost. It was an exceptional moment in Einstein's thought, which found itself changing its point of view, asking if after all such an idea, a priori so necessary, so well established, was really indispensable. Would it be possible to do without it? We often find in his work such a process of critical thinking, such a shift in the point of view, such a turn around. He had a very characteristic way of looking for an answer to questions: by turning the problem around, he evacuated it of its sense. And if there were no problem? This was how he would deal with the question of simultaneity and later with that of absolute space: there is no simultaneity, any more than there is absolute space.

Special relativity was of essential significance for Einstein's thought, for it was for him a fundamental experience that marked his style of work. In the future, he would become that man of principle, that man with a critical eye, a supporter of general but also minimal ideas. Physics must be simple.

CHAPTER 2

C: A CONSTANT OF STRUCTURE

Let us look at the question in a different way: by beginning at the end. Special relativity is a kinematics, a theory of space-time, of the *physical* relations between space and time; a theory that imposes *c* as a new element in physics. I will repeatedly emphasize the role of *c*. However, the importance of *c* viewed as the speed of light should not be exaggerated: in the equations of kinematics, *c* appears less as the speed of light in a vacuum than as the *limiting* speed of any signal.[4] This limiting speed only coincides with the speed of light insofar as a photon has zero mass. In short, *c* is rather a theoretical constant of structure of space-time and it is related to the question of causality.

Thus, *c* is an essential element, a primal, fundamental one in physics. Today, this appears obvious but at the time it was shocking. And yet, physicists should have long suspected that light had something to do with kinematics and causality—since the end of the seventeenth century, after Römer, for instance. The story of Römer and his measurement of the speed of light, which was related earlier, is based on a similar idea, for it was not an instantaneous measurement that Römer took, but rather a sequence of measurements: Römer was "filming." I will explain myself or, rather, I will revisit some features of Römer's analysis. What is measured is certainly not the precise *moment* of the emersion of Jupiter's satellite but the moment of the *appearance* of this emersion. What we receive is an image that has aged, and whose age depends on the time that light has taken to reach us. We are therefore dealing here with the successive *appearances* of the emersions, and we receive the distorted film of the satellite's emersions, slowed down or accelerated depending on the motion, like the film of a periodic motion taken from a moving train. In fact, the phenomenon is no different from that of the varying frequency of a locomotive's whistle as it passes through a station, and whose sound changes as a result of the motion of the source with respect to the observer on the platform; the appearance of the emersions is subject to a Doppler effect.

Figure 2.1. Andromeda, the most distant object visible to the naked eye. (T. R. Lauer [NOAO], NASA)

It is an ordinary, quite common phenomenon, and it is simply a matter of reflecting on it. The image we see, the snapshot of a scene that is formed on the film or on the retina, are only instantaneous at the moment the shutter (or the photographer's conscience) closes. The foreground, middle distance, and background of the scene whose picture we take—my daughter and her cat, a tree, the Moon, some stars—are certainly taken at the same moment, but the ages of their images are not the same. The picture is therefore the superposition of the waves we have just caught at the same instant on our film; but even if they have been captured at the same instant, they have not been born at the same time, and on the picture, the age of each of those persons is not their true age. On the observatory's negative, as well as on our family photographs, only the trigger is instantaneous; all the rest—foreground, background, and the starry sky—are not really seen at the same instant. There is a difference of 10^{-10} seconds (one-tenth of a billionth of a second) between that person's face and the cat on the sofa. The image of the Moon takes one full second to reach us, that of the Sun between eight and nine minutes, and that of the nearest star at least four years. The star lies about its age by several years; it sends us pictures of its youth. The light from some stars only reaches us after billions of years. Where is the instantaneous picture?

In fact, *c* is everywhere and simultaneity is an illusion. A snapshot is composed of a stacking up of the various perspectives; nothing, apart from the photograph itself—but excluding the events it depicts—is simultaneous or instantaneous. On our retinas, on the emulsion coating of our film, events stack up and superpose while appearing to be concurrent. The image is a superposition of thin layers, the various levels of depth of the scene, and by slicing it up we can recover the different temporal layers. And that is precisely what Römer did in order to analyze Cassini's tables: to understand the motion of Jupiter's moons he had to take into account the speed of light, the speed of transmission of the signal he was using.

If we resort to light, it is not by accident but because light is part of the process that allows us to see and hence to measure. We *see* the ball rolling inside the moving train. Yet there is another scoundrel playing a fundamental role in what we see: the light that has conveyed the signal. Historically, the question arises in an astronomical context, where the ball is replaced by a light particle and the train by the Earth in motion, with the astronomer playing the role of stationmaster. In the same way, it is not possible to define the simultaneity of two distant events without employing a signal, and the signal of choice, historically, is the fastest known signal: a light signal.

The finiteness of the speed of light has many consequences of which we are not immediately aware, so ingrained is our habit to take for granted the instantaneous character of what is "seen." And it is far from certain that after Römer's discovery all natural philosophers clearly understood that what they observed was only the image of something that had taken place some time (a long time, in certain cases) ago. The consequences of Römer's discovery were extremely important for our vision of the world from a philosophical point of view, among others, but it appears that at the time this was not fully appreciated. The realization of these consequences would take place only gradually, for it was at the end of the nineteenth century, if not at the beginning of the twentieth, that all the implications for kinematics were drawn.

And yet, at the very beginning of the eighteenth century an English natural philosopher, William Whiston, drew an analogy—a very telling analogy—between the time it took for the sound from a battle to reach a distant observer and the time taken by light to come down from the sky: "For in like manner, as we at length hear of Things done by Sea or Land, Battles of Armies, and the like, after a competent Space of Time interpos'd, for the Passage of the Messengers; so the Things done in the Heavens, and which are from thence brought to our Eyes by the means of the Rays of Light, as certain Ethereal Messengers, cannot come to us, or be seen by us, until some time after they are done; and this in proportion to their Distance."[5]

And so, when we listen to a scene, we hear the various levels of reality, the depth and width of the sounds. Each sound lives in space and in time, in space-time: it has a place and a time. Thanks to this analogy we can more easily accept the unacceptable: that we can see something that no longer exists. It is a useful remark, one that allows us to better understand, to better *hear* the role of light in kinematics.

In the preface of his 1802 catalogue, William Herschel was fully aware of that:[6] "when we see an object . . . one of these very remote nebulae . . . , the rays of light which convey its image to the eye, must have been more than nineteen hundred and ten thousand, that is, almost two millions of years on their way; and that, consequently, so many years ago, this object must already have had an existence in the sidereal heavens, in order to send out those rays by which we now perceive it."[7]

And to a poet who was paying him a visit in 1813 (Herschel was then 75), he said: "I have looked farther into space than ever human being did before me. I have observed stars of which the light, it can be proved, must take two million years to reach the earth. Nay more, if those distant bodies had ceased to exist millions of years ago, we should still see them, as the light did travel after the body was gone."[8]

And so, already, *c* is *clearly* essential for any precise description of even moderately distant objects. Today—but only recently—it is a fashionable subject, for thanks to cosmology those are now familiar ideas. The big bang, the date and place of the birth of our universe, is about fifteen billion years in the past, and distances to nebulae are expressed in *light-years*, a revealing term showing that astronomers regard space and time as truly connected. We hear about the immensity of the cosmos, about the immensity of the time that light takes to travel from stars and galaxies, about the images used to interpret all this: a photograph of the sky, a strange image of the world, a piling up of planes (in fact, of spheres) of time and events that land on our retinas since the beginning of the world.

But the real novelty of "special" relativity is the presence of *c* in the equations of kinematics and, through kinematics, in the whole of physics. Thus, from now on, *c* becomes not so much a new constant of physics as a constant of space-time structure.

SIMULTANEITY OR HOW TO BE EVERYWHERE AT ONCE

Lorentz naturally considered simultaneity a primitive concept requiring no discussion, and he was neither the only nor the last one to think so. And who could blame him? Wasn't it one of the most deeply rooted certainties in our vision of the world? Was there a most obvious concept than simultaneity? Simultaneity at a distance had a meaning that no one would call into question. It was plain common sense. I would even go as far as to suggest that simultaneity is a notion more fundamental, more ancient, and more evident than absolute time, which is a relatively involved idea whose definition depends, in fact, on that of simultaneity. Conversely, once Newton's absolute and universal time ruled over the order of the world and events, simultaneity followed. And so we believe that we can unambiguously tell whether two events, one taking place in Bern, in Zurich, or at the Empire State building in New York, and the other on a train traveling to Detroit, on the Moon, or in the Scorpio nebulae, are simultaneous or not. This is true up to a certain point; if we receive a telephone call, the time lag—which we can now clearly hear if the connection is made via satellite—will be a telling factor. But we can never know (even if we see him) what our caller is precisely doing at a given moment: we always see others in the past.

From a classical standpoint, time and space were concepts that preceded all experience, all knowledge; they could not be, it appeared, the subject of any experiment or definition. Neither space nor time could be called into question or experimented on. Which philosophy did not rank simultaneity at the top of its certainties? Isn't it still today everyone's natural philosophy? In the *Definitions* of his *Principia*, Newton wrote: "As the order of the parts of time is immutable, so also is the order of the parts of space. . . . All things are placed in time as to order of succession; and in space as in order of situation. It is from their essence or nature that they are placed."[9]

And simultaneity as such is not even stated—it is inherent in the order of succession, in the very order of things. At the very most Kant, in *Critique of Pure Reason*, raised the question of the representation of time as an a priori concept that implies that of simultaneity, and hence as something that is given.

With a clock in his hand, Newton's God oversaw, then, the harmony of the universe; it was a conspicuous clock, which could be seen from everywhere. God's look was immediate, and He continuously had an instantaneous and complete image of the world. Everybody was supposed to be able to read Newton's clock directly, immediately, and without giving it a thought.

How could anyone have dared raise the question of the speed of God's look? Heathen! From a physical point of view, it all worked rather well. For it was God, and God alone, who could be everywhere at the same time. In the impossibility of such ubiquity, of such omnipresence, lie the foundations of relativity.

The problems appeared in the early twentieth century with electrodynamics. Measurements became so accurate that difficult problems arose which could not be worked out. As it became clear later, those problems were related to the speed of transmission of signals. And it took time to understand, to accept that ordinary time—that time on which one could count, that implacable time—should be questioned. The time it took to figure out time? One century! Is it at all surprising that figuring out time should be so difficult?

Simultaneity is a delicate notion, one that is a little too obvious, one to which we have become too accustomed. It is not so much a question of understanding as it is a question of accepting that there is nothing to understand: that there is no absolute simultaneity, no absolute time, and no problem. There is no problem, indeed. First, it is enough to understand that the problem can be solved by means of a pretty nice pirouette. Then it will be necessary to *really* understand, that is, to seize upon the new tools, the new concepts, and the new spaces and to forget the old ideas, which are always much too apparent, much too close. All of that cannot be achieved without a task of reconstruction, a task involving a revamping of concepts and equations.

In fact, only the simultaneity of two concomitant events—essentially a collision—retains a physical meaning. If you bump into your cousin on a street corner, then both of you know that the collision took place at the same time and at the same place. It is even one of the few absolutely indisputable concepts in general relativity. But simultaneity at a distance is a different thing, because then a signal is needed to convey the information from one point to another; a signal with a speed that has to be measured and defined. But how? This requires clocks and additional hypotheses and a definition of simultaneity at a distance.

POINCARÉ: FROM LORENTZ TO EINSTEIN

Before discussing the birth of special relativity (which is usually a bit too hastily credited to Einstein alone), let us take a look at the work of the French mathematical physicist Henri Poincaré, who studied these questions in great detail. Many ideas, criticisms, and concepts were shared by both men.[10] It is possible that some similar analysis led them to the same conclusions or that Einstein seized upon certain ideas of his senior. In any case, it is difficult to put into perspective the influences that he might have had, all the more so since documents from this period are very scarce.[11]

As a student, Einstein had read Poincaré's *La Science et l'Hypothèse* (*Science and Hypothesis*),[12] one of the very first books (together with Mach's *Mechanics*) that he included in the program of the Olympia academy which, from 1902 to 1904, brought together three young men with a passion for physics and the philosophy of physics and who gathered to read, discuss, and dine. The other two members of the group were Maurice Solovine, who met Einstein through an ad where the latter offered "physics lessons at three francs an hour,"[13] and Conrad Habicht, a friend of Einstein's. Both men were witnesses at the wedding of Albert and Mileva in January 1903.

La Science et l'Hypothèse was published in 1902, and the ideas discussed by Poincaré in it are precisely those in which Einstein was interested. (In 1956, in his introduction to the publication of the letters he received from Einstein, Solovine would remember "that book which so deeply impressed us and fascinated us for weeks."[14]) This is not surprising, for Poincaré was an expert in those subjects, which he developed from a very original angle. There is no doubt that such a book, which seemed almost to have been written for him, had a profound influence on Einstein.

Naturally, Poincaré's interest in those subjects preceded that of Einstein's, because the French mathematician belonged to Lorentz's generation. Already in 1891 he had published an article on non-Euclidean geometry[15] and another one in 1898, "The

Measurement of Time,"[16] in which he drew attention to the conventional character of the definition of simultaneity for distant events, and in 1900, he proposed a physical interpretation of Lorentz's local time.[17] All of these topics reappeared in his 1902 book. In *La Science et l'Hypothèse*, Poincaré speculates about subjects that would become—had, in fact, already become—also Einstein's: absolute space, the relativity of motion, duration, simultaneity, and Euclidian geometry. In order to judge for ourselves, here is an excerpt from the book, an introduction to the chapter on classical mechanics:

> 1. There is no absolute space, and we only conceive relative motions; nevertheless, facts in mechanics are usually stated as though an absolute space existed, and to which they could be referred.
> 2. There is no absolute time; the statement that two events have the same duration is in itself meaningless and it can only acquire a meaning by convention.
> 3. Not only do we not have a direct intuition of the equality of two durations but we do not even have that of the simultaneity of two events taking place at different theaters . . . .
> 4. Finally, our Euclidean geometry is only a kind of linguistic convention; we could state mechanical facts with reference to a non-Euclidean space which would be a less convenient—but equally legitimate—frame of reference than our ordinary space; such a statement would then be much more complex but it would still be possible.
>
> Thus, absolute space, absolute time, and geometry itself are not required conditions for mechanics; all those things do not preexist mechanics any more than the French language logically preexists the truths that are expressed in French.[18]

We should notice, however, that Poincaré ends this introduction by "provisionally" assuming "absolute time and Euclidean geometry." This already suggests a hesitation, one that the young Einstein would not share and which would enable him to take advantage of the new ideas without being ensnared by metaphysical doubts.

But Poincaré did not stop at the above considerations. Many of his articles contained advances in various directions that would prompt more than one author to consider him the father of special relativity, such as the English physicist Sir Edmund Whittaker, a famous expert in relativity who in his scholarly book *A History of the Theories of Aether and Electricity* called the chapter devoted to that topic "The Relativity Theory of Poincaré and Lorentz."[19] Here Whittaker clearly showed his colors: Einstein is conspicuously absent. I shall not elaborate on this dispute, which has been long argued but not settled.[20]

In 1905, Poincaré finished two papers on the dynamics of the electron. The first one was presented to the Paris Academy of Sciences at its 5 June meeting and then appeared in the *Comptes Rendus;*[21] the second one was submitted on 23 July to an obscure Italian journal, *Acts of the Palermo Mathematical Circle*, and was published in 1906.[22] It is clear that Einstein could have not read any of them before submitting, on 30 June 1905, his own article announcing special relativity. And yet, how relevant and interesting he would have found them! For Poincaré mentions the negative result of Michelson's experiment and concludes that "this impossibility to prove absolute motion appears to be a general law of nature."[23] He then sets out to demonstrate that the transformations of the electromagnetic field equations, which he calls "the Lorentz transformations," form a group.[24] But there is more: he closes his article by wondering about the effect of the Lorentz transformations on forces in general and in particular on gravitation. It is a question that Einstein would not ask until 1907 and which would lead him to general relativity. In that same article, Poincaré assumes that gravitation propagates "at the speed of light" and introduces the idea of a "gravitational wave." In his 1906 paper, the Frenchman lays the foundations of a theory of gravitation that, in his own words, "would not be altered under the group of Lorentz transformations,"[25] that is to say, it would be "Lorentz invariant." Moreover, in the same article Poincaré introduces time as a fourth imaginary coordinate as well as the four-dimensional approach,[26] which Minkowski would make precise in 1908 without mentioning Poincaré.[27] As for the latter, until his death in 1912 none of his publications would mention Einstein's theory of relativity. In his article on the origins of special relativity, Holton sees Poincaré as procrastinating, even backing away from, instead of fully grasping or achieving, the great renewal that was implicit in his work.[28]

CHAPTER 2

SPECIAL RELATIVITY PREVAILS

In May 1905, Einstein wrote to his friend Conrad Habicht to ask him for a copy of his dissertation:

> Don't you know that I am one of the 1½ fellows who would read it with interest and pleasure, you wretched man? I promise you four papers in return, the first of which I might send you soon, since I will soon get the complimentary reprints.[29] The paper deals with radiation and the energy properties of light and is very revolutionary, as you will see if you send me your work *first*. The second paper is a determination of the true sizes of atoms from the diffusion and the viscosity of dilute solutions of neutral substances.[30] The third proves that, on the assumption of the molecular theory of heat, bodies on the order of magnitude 1/1000 mm, suspended in liquids, must already perform an observable random motion that is produced by thermal motion;[31] . . . The fourth paper is only a rough draft at this point,[32] and is an electrodynamics of moving bodies which employs a modification of the theory of space and time; the purely kinematics part of this paper will surely interest you.[33]

Four papers, three of which would revolutionize their respective fields and each one of them worthy of a Nobel Prize—recognition, by the way, that would not be granted to their author for a long time yet. Does the label "miraculous year" for Einstein, when referring to 1905, need further justification?

However, it is striking to realize the degree to which a critical perspective pervades the article introducing special relativity, an expression that would appear only later. It was an essential trait of Einstein's character, of his *style*. He liked to call into question, to ask questions—impertinent questions, preferably—to criticize, to turn the point of view upside down:

"But what do you mean, my dear colleague? What do you really mean by that?"

"Come on," replies the other, "obviously . . ."

Obviously—the big word is out; Albert is at ease. " 'Obviously,' 'obviously,' you have said 'obviously'? And yet . . ."

Einstein smiles and rubs his hands in anticipation. He is in his element: finding flaws in concepts that are too *obvious* will be his specialty. "Is it really that obvious? But what does it mean, then?"

". . ."

"And if we looked at it in a different way?"

We have seen an example of this in his criticism of the double description of the same phenomenon, of this conceptual asymmetry, that opens his article on special relativity. He immediately adds another objection regarding the notion of absolute rest, to which does not correspond any property of phenomena, and his target is of course the ether. What is it good for? What's the point of keeping a useless concept? In the next page he calls into question yet another concept: the definition of absolute simultaneity. Einstein has his own characteristic way of solving very delicate questions: he eliminates them. Simultaneity causes a problem? But what does it mean? Do we really need it? And what if we just got rid of it?

And so Einstein carried on, dismantling concepts. Nevertheless, one needed to know of what (and of when) one is talking about and then combine all that into a consistent whole. It would be the task of special relativity to ensure the consistency of those distant events by giving an intrinsic description of them while at the same time preserving causality.[34]

It is an amazing article, hardly technical, and certainly not a scholarly one. The mathematics employed is completely basic, and there are no references to other papers or to scientific experiments. But there are thought experiments, and the article contains certain naïve remarks that no doubt annoyed some of his colleagues, for example, his explanation of what he meant by simultaneity. It is a linguist speaking, analyzing, disassembling time the way one would take apart the pieces of a clock or a puzzle. He dares to clarify the meaning of the sentence "that train arrives here at 7 o'clock," which, he observes, means that "the pointing of the small hand of my clock to 7 and the arrival of the train are simultaneous events." Amazing. Many of his distinguished colleagues must have said to themselves: "This young fellow, does he believe he needs to tell us what time it is? What a nerve! Really! How can a serious journal accept such naïveté; *Annalen des Physik* is no longer what it used to be! But how on earth could they accept such a paper!"

Such then were the questions that led Einstein to believe that he had to start from scratch, and he loved doing that; he was a slightly revolutionary young man. And if we did away with all that jumble? He showed that it was entirely possible.

After all these highly critical considerations, Einstein has to rebuild everything and lay down the bases of kinematics, of physics, from scratch. He states two principles: the principle of relativity, which establishes the invariance of the laws of physics in every inertial frame (see box 2), and the principle of the constancy of the speed of light—two principles, the first of which was already well known because it was Galileo's principle, which was until then restricted to the mechanics of material bodies but now also applied to optics and electromagnetism. As for the constancy of the speed of light, it was not really a new experimental fact, since after Bradley and aberration, following the experiments of Arago, Fizeau, Michelson, and Morley, this constancy was implicitly accepted; the novelty consisted in making it into a principle.

At the same time, that plethora of hypotheses and all sorts of tricks that Lorentz had to use to account for the electromagnetism of moving bodies disappears. Those ad hoc hypotheses are now reduced to two simple principles! Two simple principles but at what price? Einstein makes a clear-cut decision; it is of course a conceptual revolution. Only much later would Einstein try to explain his motives and what he allowed himself to do: to believe

Box 2. The Principle of Relativity and Inertial Frames
Physics must be everywhere the same. Otherwise, how could we apply its laws? But, everywhere? That is precisely the problem, and the question we should be asking is, in what context can we apply the laws of physics? The answer is (in Galilean physics as well as in special relativity): in the inertial frames. The laws of physics have the same form in all inertial frames. In other words, they are invariant with respect to all transformations from one inertial frame to another; relativity—Galilean as well as special—is the theory of those transformations.

(*continued*)

Box 2. *(continued)*

But what is an inertial frame?

Notice that so far we have considered only constant velocities: our travelers, material or light particles, and trains are supposed to have a constant velocity with respect to each other and also with respect to the frames in which they are observed and measured. An inertial frame—also called an inertial reference system—must not be confused with a coordinate system. The latter is simply a way of locating an event in space—its address, so to speak; it is generally a trihedron (the three edges of a cube) plus time. This enables us to situate in a precise way any event or object in space and in time.

The inertial frames (or reference systems) are particular frames with respect to which classical physics is built. In an inertial frame, a free particle (i.e., one that is not subject to any external force) will follow a linear path with constant velocity; this is the principle of relativity. The external forces on a given particle are defined using this principle: if the particle's trajectory is inertial, then the particle is not being subject to any force and, conversely, if its trajectory is not inertial, it is because some external force is being applied to the particle. Thus, the law of inertia is valid on . . . the inertial frames. It is the principle of relativity, with respect to which absolute space has been defined since the seventeenth century. Of course, all frames traveling with constant speed with respect to a given inertial frame, or with respect to absolute space, are themselves inertial. In classical kinematics, the transformations of Galilean coordinates allow us to go from one inertial frame to another. In relativistic kinematics that role is played by the Lorentz transformations.

In classical physics, the principle of relativity applied only to the laws of mechanics. In special relativity, this principle is extended to cover, in addition to the equations of mechanics, also those of optics and electrodynamics. It is the new, larger "group" of transformations that makes this extension possible.

that it was possible to sacrifice our most fundamental concepts and in particular those involving space and time: "Mechanically, all inertial systems are equivalent. In accordance with experience, this equivalence also extends to optics and electrodynamics. . . . I soon reached the conviction that this had its basis in a deep incompleteness of the theoretical system. The desire to discover and overcome this generated a state of psychic tension in me that, after seven years of vain searching, was resolved by relativizing the concepts of time and length."[35]

Many consequences followed from this new structure, to which he has to add a definition of simultaneity, for simultaneity is now a relative notion: two events that are simultaneous with respect to a given inertial frame need not be simultaneous with respect to another such frame.[36] Likewise, the length of a rigid ruler now becomes a relative concept: length contracts from one inertial frame to another; and the duration of phenomena also becomes relative: there is dilation of duration.

Thus, length and duration are no longer intrinsic magnitudes but, rather, depend on the frame of reference with respect to which they are measured; they are relative. Insofar as these magnitudes are not intrinsic, it will be difficult to use them correctly as physical observables. In his 1905 article Einstein still uses various concepts that are not intrinsic, that is, that are not invariant. A physicist needs intrinsic physical magnitudes defined independently of any given reference frame and which can subsequently be projected on the particular frame under consideration where they will then be observed, that is, what physicists call observables. It is precisely in that sense that relativity is an absolute, invariant theory and not a relative one: it is its expression in a particular frame that is relative. It would take many articles and other publications for special relativity to be better understood and recast. First of all, it would take Minkowski's paper—to which we shall come back—but also general relativity since, to really understand his paper, one must first proceed to general relativity in order to appreciate the difficulty of the undertaking.

In the second part of his paper, Einstein obtains, in a much simpler way, not only the set of then classical results and, foremost, the Lorentz transformations expressing the possible change of coordinates between two inertial frames, but also the relativistic version of the Doppler effect, of aberration, of the way certain physical observables (notably frequency and energy) are transformed by a change of frame, and so forth.

Einstein's article founding the new kinematics appeared on 26 September 1905 with the title "On the Electrodynamics of Moving Bodies," two months after his paper on Brownian motion (18 July) and three months after the publication of his article on the hypotheses of light quanta and the photoelectric effect (9 June). Each of the three papers deals with a different subject. They are all remarkable if not revolutionary, and any one of them would have been enough to establish Einstein's reputation. On 21 November of the same year, he published a short, three-page article, a consequence of special relativity showing the equivalence of mass and energy and introducing the famous equation $E = mc^2$. It was indeed an extraordinary year for Einstein, a "miraculous year."

As expected, the discussion, acceptance, and transmission of special relativity was not a simple formality. It was a revolutionary, an absolutely revolutionary, theory calling into question our conceptions of space, of our universe, and of the very basis used to describe what would later be called *space-time*. However, insofar as it solved—by simplifying them, by *dissolving* them—so many problems, the theory was quickly accepted, although not necessarily understood. It was not so much because Einstein convinced or seduced; he prevailed, and it was very difficult to resist him, because he proposed ideas that are simple, ideas that work. He prevailed because of the strength and the power of explanation and prediction of his ideas.

Thanks to his sister Maja, we know that after submitting his article to *Annalen der Physik* Einstein was worried for a while, fearing that the article would be rejected by the prestigious German journal. He anxiously waited for an answer. This is understandable, for Einstein was aware of the obviousness, the force, the necessity, and the difficulty of understanding and accepting the revolution he proposed.

In fact, his article immediately elicited some positive comments. In the same year, 1905, the article was cited by Walter Kaufmann in connection with his experiments on the mass of the electron, and then in 1906 by Paul Drude, the editor of *Annalen der Physik*, both in one of his own articles and in a review of the *Handbuch der Physik*. In 1906, Einstein corresponded with Max Planck, and in 1907 with Max von Laue, Hermann Minkowski, Wilhem Röntgen, and Wilhem Wien. Arnold Sommerfeld, very impressed, prepared a series of lectures, for what was probably the first course on the new theory. In 1907, Max von Laue, Planck's assistant, and Kurd von Mosengeil, one of Planck's students, published a paper on the subject, and Johannes Stark, the editor of the *Jahrbuch der Radioaktivitat und Elektronik*, asked Einstein to write a review article on special relativity.[37] In short, everything was moving along fine, and the idea generated more than simple interest.

Nevertheless, special relativity raised many delicate problems, questions whose subtlety sometimes stemmed from the difficulty of thinking in an unusual context. The Langevin twins paradox, for instance, would prove too much for many physicists! But other difficulties would only be misinterpretations of a word that became fashionable with a little too much hype.

Sixteen years later, in 1921, Einstein would be awarded the Nobel Prize for his "contributions to theoretical physics and in particular his discovery of the law of the photoelectric effect." No mention would be made of special relativity, a sign that the theory had not yet entered mainstream physics; never mind general relativity, which Einstein had formulated in the meantime.

MINKOWSKI: A NON-EUCLIDEAN SPACE

Let us go back for a moment to the classical model: three-dimensional Newtonian space with time as a parameter for motion. If we draw a diagram to represent the motion of the train with respect to the tracks, space will be represented along the x-axis and time along the y-axis. It is clear that in order to *see* the space, one must travel in it. Space will be—and we shall come back to this, in particular in general relativity—the space *reached* by some particle; except as a place that can be occupied or that is occupied, space has no meaning. That is not yet what special relativity says, but this little detour through general relativity enables us to ask the questions: Without time, what is space? Without time, does space exist? Space is always the space traveled by something. This is also what the different depth levels of a photograph, each of which is dated, show. The link between space and time is thus a familiar fact that existed before special relativity. The latter required, starting with Lorentz, an interdependency (not well understood at first) between space and time; later, thanks to Einstein, a definition of physical time tied to the motion of the observer; and finally, thanks to Hermann Minkowski, it called for *space-time* as a single concept, which in this relativistic century has become *spacetime*—without a hyphen.

Einstein's presentation of his special relativity, the point of view he had proposed in his earlier articles, concealed its revolutionary character behind physical developments that were essentially classical and had a strong epistemological flavor. In 1908, Minkowski proposed a radically new vision of the theory, an analysis that Einstein found difficult to recognize and even to "digest," but one which would soon have a decisive influence on his own work. Without Minkowski's interpretation Einstein would not have been able to conceive or to write his general relativity theory so effortlessly.

In the introduction of the talk he gave in Cologne in September 1908, and which would later become a classic of relativistic literature,[38] Minkowski explicitly presented relativity as a revolutionary theory: "The views of space and time which I wish to lay before you have sprung from the soil of experimental physics, and therein lies their strength. They are radical. Henceforth space by itself, and time by itself, are doomed to fade away into mere shadows, and only a kind of union of the two will preserve an independent reality."[39]

This radical view was absent from Einstein's analyses and presentations, a fact that Minkowski did not fail to notice in his talk, while making a barbed remark about his former student: "Neither Einstein nor Lorentz made any attack on the concept of space," he said, before adding: "One may expect to find a corresponding violation of the concept of space appraised as another act of audacity on the part of the higher mathematics."[40]

Minkowski showed that three-dimensional space was now outdated: "The objects of our perception invariably include places and time in combination. Nobody has ever noticed a place except at a time, or a time except at a place."[41] While declaring, "But I still respect the dogma that both space and time have independent significance,"[42] Minkowski set up a four-dimensional version of Einstein's relativity that had a tremendous influence on the evolution of the theory.

He thus introduced four-dimensional objects and concepts that would play a fundamental role. He replaced the notion of a point by that of a universe point (no longer only three space coordinates but three space coordinates plus one time coordinate), and that of a curve by that of a universe line (no longer curves to be traveled, or *geometric curves*, but curves that had been traveled, or *trajectories*). Those universe points would later be called events, but the concepts would remain. It was simply a matter of conforming to perception and no longer speaking of space and time separately but, rather, together: we only observe events. A universe point, an event, is a place-at-a-given-moment, while a universe line is a set of universe points, a trajectory in space-time. But, above all, Minkowski introduced relativity's fundamental tool: *proper* time, which he constructed from the line element or *metric* of space-time. To sum up, Einstein had not pushed his analysis to its ultimate consequences.

As we have seen in chapter 1, the classical, linear law of the addition of velocities follows from some strong but quite natural hypotheses: Newtonian space as a framework, that is, an absolute space of three dimensions endowed with an absolute time, and a definition of speed that is "taken for granted" (distance divided by time).

FROM CLASSICAL DISTANCE TO PROPER TIME

Now, if experience had shown that the classical law of the addition of velocities was not right (and that is exactly what happened), a first response, specifically, Lorentz's, consisted, without calling into question classical space and time, in imagining a contraction of lengths and a dilation of durations in order to account for the new facts. Einstein, for his part, challenged those hypotheses, that is, the classical, absolute character of the definition of space, time, and speed. Special relativity resulted from this analysis, and it involves, as Minkowski has shown, a space that is no longer Newtonian but a different, really different, one now known as Minkowskian and in which there is neither absolute distance nor absolute time anymore. Such is the strange space that has served from then on as the model: a space whose fundamental structure, whose topology (in short, the science of shapes and geometric structures not involving distance) is far removed from that of Euclidean space. This is a consequence of the existence of a limiting speed, from which follows that certain events cannot be causally linked. Similarly, the non-Euclidean nature of the representation space is at the root of certain difficulties we have encountered and which will reappear later.

Classical distance and classical time have to be abandoned. Let us give one more reason why the notion of a rigid, solid body, which is another classical notion, must also be abandoned. The crucial point here is that an absolutely rigid body should instantaneously transmit a signal from one end to another, and hence the speed of propagation of that signal would be infinite. But something is not right, since the new kinematics prevents the existence of speeds that exceed that of light. Now, a solid is just a nondeformable body of a fixed, invariant length. But the length of our solid is not an invariant, a fact that follows from the contraction of lengths. Thus, concepts such as classical distance, solid, rigid body, and classical time are incompatible with relativity; we shall have to do without.

A fundamental question then crops up: what will take the place of distance and time? Since it is subject to contraction, coordinate distance can no longer be an invariant, any more than coordinate time, which dilates. None of them has a physical sense anymore; they are not physical observables. No invariant, no intrinsic magnitude can give them *separately* a meaning. But there exists a magnitude combining time and distance (and—surprise, surprise—related to the speed of light) that is conserved in all inertial frames: proper time.

Proper time is not a universal time, such as Newton's, which is valid throughout the whole universe, but, rather, proper time is a time that is defined only locally. Its definition is, of course, universal, but it remains local and must be adjusted to each particle, each observer moving around in the universe. It is a time that must be, so to speak, computed over and over again, integrated along each trajectory. In short, every thing has its time—its proper time.

However, the mathematical structure, the definition, of that proper time was not already available, so it had to be created. What is then the nature of proper time? It is clear that it will be a clever combination of coordinate time and coordinate distance; but how exactly is one to construct it? What is its mathematical structure?

Let us warn the reader that its expression is not simple or obvious. It is not a "global" analytical expression directly involving coordinate distance and coordinate time. It is, rather, an infinitesimal, differential expression defining the interval of time between two very close events that will have to be integrated along the trajectory in space-time between the events whose difference in proper time, or age, we wish to obtain. From that calculation it immediately follows that the age of a traveler—one of the Langevin twins, for example—or the life span of an elementary particle will depend on its trajectory. This is perfectly logical, for proper time replaces classical distance, which we know does depend on the path traveled.

In the Euclidean framework, metric distance is expressed, using Pythagoras's theorem, as a sum of squares (see box 3). And so, the distance between any two points is always positive; it can only be zero if these are one and the same point.

Box 3. "Metric" and Proper Time

In special or general relativity, the definition of proper time is based on that of distance in a "metric" space—in Euclidean space, for instance. In order to measure the length *ds* of a vector dM in the Euclidean plane, it is enough to know the projections *dx*, *dy* on each of the coordinate axes. Then, the "infinitesimal distance element" or "metric" is simply the square of the length of that vector, which is written, thanks to Pythagoras's theorem:

$$ds^2 = dx^2 + dy^2.$$

We then have to integrate that infinitesimal vector, *ds*, along the curve "P_0P" whose length we wish to measure:

$$s = \int_{P_0}^{P} ds = \int_{P_0}^{P} \sqrt{dx^2 + dy^2}.$$

On the sphere, the form of the line element of distance must take into account the use of polar coordinates (θ, ϕ) as well as the curvature of the sphere, so that "ds^2" is written:

$$ds^2 = r^2(d\theta^2 + \sin^2\theta d\phi^2)$$

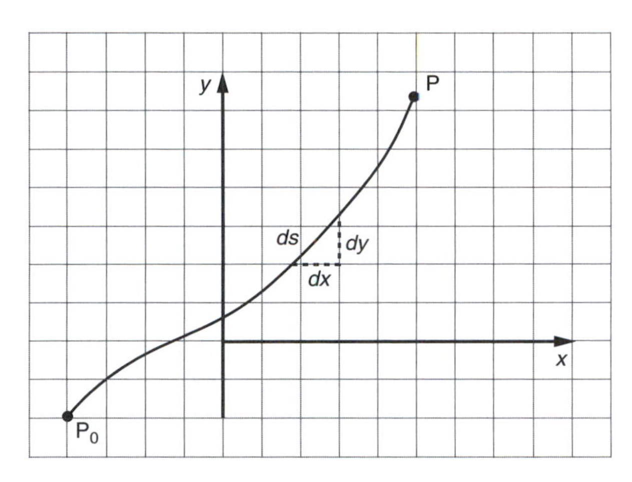

Figure 2.2. The measure of the length of a plane curve.

(*continued*)

Box 3. *(continued)*

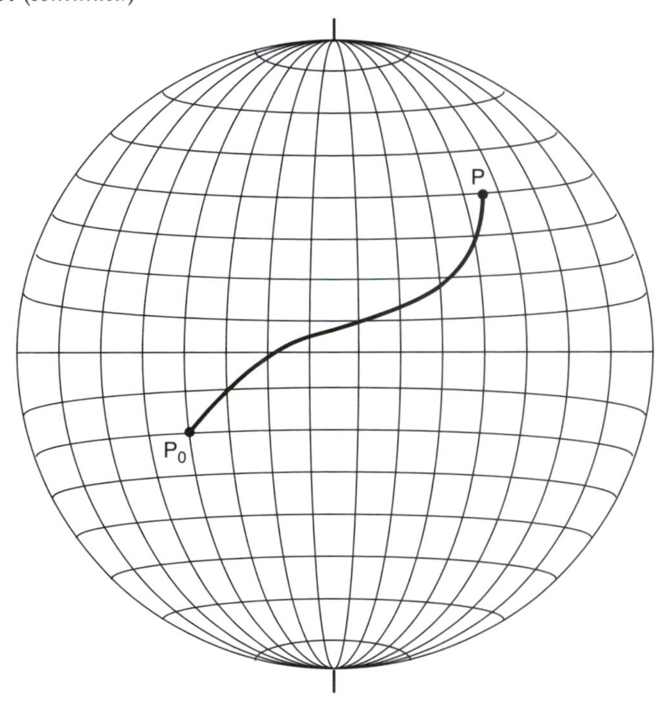

Figure 2.3. The measure of the length of a curve drawn on a sphere.

In Minkowski's space, the line element of space-time, ds (which is in fact the infinitesimal element of proper time) takes the form:

$$ds^2 = c^2dt^2 - (dx^2 + dy^2 + dz^2),$$

which can also be written in polar coordinates as follows:

$$ds^2 = c^2dt^2 - dr^2 - r^2(d\theta^2 + \sin^2\theta d\phi^2).$$

Thus, in special relativity, the proper time s between two events P_0 and P on the universe line will be the integral of ds in Minkowski space:

$$s = \int_{P_0}^{P} ds = \int_{P_0}^{P} \sqrt{c^2dt^2 - (dx^2 + dy^2 + dz^2)}.$$

If we no longer require the above property of distance to hold, the situation gets complicated. This is the case of the "space-time" structures we are interested in, in particular Minkowski space-time, whose metric is defined by

$$ds^2 = c^2dt^2 - (dx^2 + dy^2 + dz^2).$$

The fact that the last three coefficients are negative does not result in a more complex situation. The time coefficient is c^2 to insure that the metric is positive for material particles, whose instantaneous speed is smaller than that of light.

Nevertheless, one may wonder what the origin of this strange expression is. To begin with, it generalizes Euclidean distance, but the real reason for the particular form of the above formula is that proper time, and hence ds^2, must be invariant under the Lorentz transformations, which is the case for this expression. As Minkowski himself put it: "The differential equation for the propagation of light in empty space possesses that group [of transformations]."[43]

Physical laws and magnitudes must be everywhere the same; everywhere, that is, in every inertial frame (see box 2). This is of course the case for proper time, the most important physical observable in relativity, which must therefore be *invariant*. But invariant under which transformations? Under those transformations from a given inertial frame to another, naturally.[44] And relativity, Galilean as well as special, is the theory of those transformations—Galilean in a classical setting, or Lorentz in a relativistic one—that permit the transition from one inertial frame to another.

The transformations that leave invariant the classical laws and in particular Newton's theories are therefore the Galilean transformations: the rotations of space and all translations in space and in time. These transformations form a group known as the "Galileo group." This means that all classical laws of physics expressed in a given inertial frame retain the same form under the action of a Galilean transformation. The same is true in special relativity, except that the rotations in question are more general: they are those of space-time, and their associated group of transformations is the Lorentz group, of which the Poincaré group is an extension that includes the group of translations. Hence, proper time is invariant with respect to the Lorentz transformations. It is a physical magnitude, an intrinsic magnitude that will also appear, with a new definition and by a suitable generalization, in general relativity; it is the keystone of relativity.

The proper time of an observer, the time displayed by his or her wristwatch, is one of the few physical magnitudes that will retain an intrinsic meaning in general relativity; it is undoubtedly the most important concept in relativity. Each particle will then have its own proper time, measured by its associated clock, but there will be no absolute time that would permit defining the simultaneity of two distinct events.

UNIVERSE LINES AND LIGHT CONES

The trajectory of a particle is described by four equations expressing its position by means of its four coordinates as a function of its proper time. Oddly, the graphical representation of such a trajectory is far from obvious; for example, the page of this book is Euclidean, whereas in Minkowski's space-time, it is a "hyperplane." The representation of this structure poses many problems. Because of their different metrics, Minkowski's space-time and Euclidean space have completely different structures, and an accurate graphical (that is, planar or two-dimensional) representation will not be possible. All diagrams of space-time will bear the mark of this difference in structure, of this difference in topology. Of course, just as in classical physics, an event will still be represented by four coordinates, three spatial and one temporal, but the analogy ends there. In fact, even the representation of physical space poses a problem, simply because the page of a book is two-dimensional. Usually, time is represented along the y-axis, so that only one dimension is left to represent space along the x-axis, which is not really a problem.

In a Galilean context, speed being unbounded, nothing prevents a particle from accelerating forever and so from being able

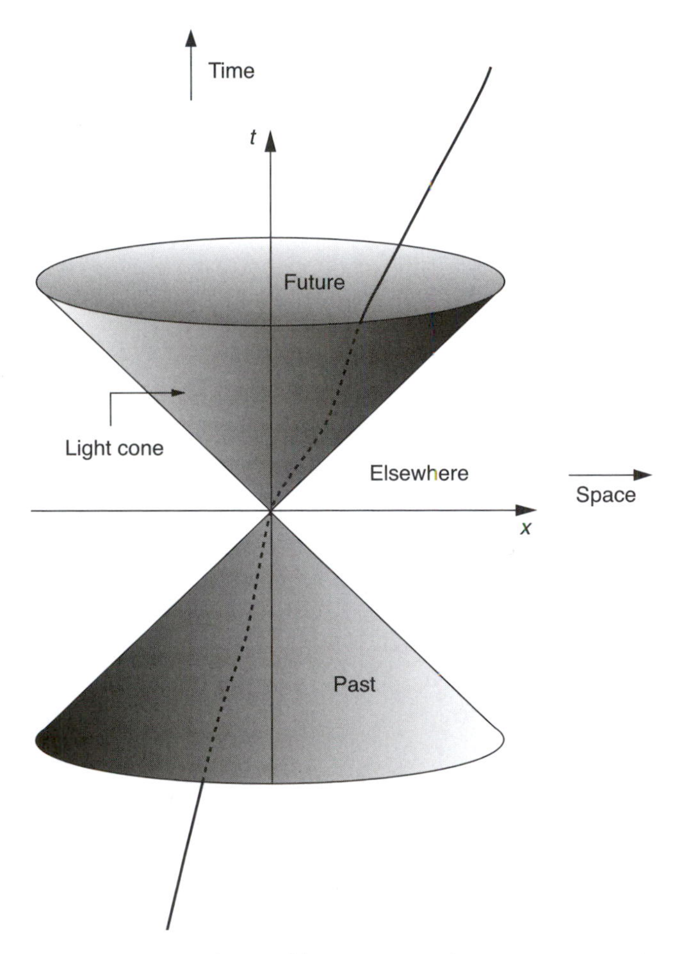

Figure 2.4. In classical kinematics, the entire space-time may be reached from any given point, here the origin, provided we travel fast enough. In relativistic kinematics, on the other hand, the "elsewhere" is forbidden, the physical space-time is bounded by the light cone.

to reach any given point of space-time—by traveling with a speed higher than that of light, if necessary. But in the new kinematics, not all points of space-time can be reached from a given point: only those such that the speed required to get there is smaller than the speed of light. In other words, in classical kinematics all of space can be reached from a given initial point provided the particle travels fast enough, while in relativistic kinematics only those points lying inside (or on the surface of) the *light cone* are accessible. In special relativity, physical space is therefore truncated, restricted by a boundary in the shape of a cone of light.

Here is an image that might help the reader to visualize what "a cone" is. You are traveling in a car whose speed is obviously bounded. Loosely speaking, your "cone" (more precisely, the interior of your cone), is the set of your possibilities, of the events (places at a given time) you can reach—or that you may have reached in the past—given your vehicle's speed limit. The interior of the cone is the set of accessible events; its exterior is the set of those events that are inaccessible, while the surface of the cone consists of the limit of the accessible events, that is, those events that you can *just barely* attain.

The cone is simply a consequence of the existence of a speed limit: c, in the case of the new kinematics (and 70 miles per hour on the highway!). Of course, a cone is always linked to an event, or point in space-time: the event associated with the starting point in space and moment in time. The set of accessible events depends on that initial event.

A light particle or photon will therefore propagate "on the surface of the cone," so that the cone is composed of photons whose time has in a sense stopped. A photon, which has an infinite life span, does not notice the time going by. Outside the cone (the nonshaded part of figure 2.4) speeds are higher than c; it is the *elsewhere*, a zone inaccessible from the initial event.

PROPER TIME

Minkowski's proper time, combining as it does time and distance, turns out to be absolutely essential in relativity. Proper time, denoted by s, is simply the time that separates two events of the same trajectory in space-time. But it has neither the properties of ordinary distance nor those of Newton's absolute time; in particular, it is not always strictly positive for two distinct events. How strange! Thus, on the surface of the cone, the proper time between two events that a light signal—or, more precisely, a particle of zero mass—can join will be zero. Proper time is the physical time of a particle that travels in space. It is in a way the true time, but a time that is no longer universal. It is the time that the clock you carry with you, your wristwatch, indicates and which, in physics, is the closest thing to biological time. As for distance, we must abandon it and make do with the proper time that an observer—or a particle—will take to travel a given path. In fact, in order to understand it, it is enough to think in astronomical terms. When speaking, even in everyday language, of the distance to a given star, we are by now used to saying, for example, that *Proxima*, in the constellation of Centaur and the star closest to the Sun, is at 4.22 light-years from Earth. What does that mean? It means that if I send a light signal that the star will reflect, it will take 8.44 light-years for the signal to make the roundtrip (I prefer to speak of a roundtrip in order to avoid problems of simultaneity involving two distant places). This is the time that light takes to come back from the star as seen by the observer. It is possible to convert this time into kilometers, if one so desires, but light-years is the way to express a distance in as neat a manner as possible. Why convert it into kilometers? A kilometer has a precise meaning on Earth, defined by the standard meter in platiniridium kept at the Sèvres Pavilion, in France. But how to make those kilometers ride a light beam? Does it make sense? Especially if we consider that the standard meter is made of perfectly rigid, nondeformable platinum, while in relativity, as we have seen, rigidity is meaningless. So why not do without all that, and in particular without classical distance, which must be rigid or lose all meaning? Let us rather use time (proper time!), as it has long been the case. Even on Earth—*qui est si jolie* (which is so beautiful), as the French poet Jacques Prévert wrote—we are used to saying "I live five minutes from the Eiffel Tower," implying minutes-by-foot or minutes-by-car, whatever the case, or even minutes-by-light, for someone living on a satellite. Ironically, it is precisely the astronomers, who had

Figure 2.5. A device for determining the value of one meter in light wave length. (Photograph Observatoire de Paris)

long replaced distance by time—more precisely, by light-years—in their map of the universe, who were most reluctant to give up their preserve and embrace general relativity, a theory that had been, after all, created for them.

Thus, only "proper time" remains; that is, the particular time of each particle, which is tied to the history of each particle. Proper time will be the physical time, the central concept of each relativity theory, the essential intrinsic physical magnitude. Unfortunately, proper time has neither all the characteristics of ordinary time nor those of classical distance; it is, rather, a clever mixture of time and distance, as space-time demands. For a given particle, it is the (true) time. If two particles follow different paths in space-time to travel from one place (in space-time) to another, they will not necessarily take the same "proper time" to make the trip along their respective trajectories. We can recognize here a property of distance: if two travelers take different roads to go from one point to another, they will not necessarily arrive at the same time, even if they travel at the same speed. That is the reason why the ages of two twins who leave from the same place and meet again after having traveled different paths may not be the same when measured in their respective proper times. The fact that "proper time" is not a "coordinate time" poses a serious problem because it cannot, in general, be defined everywhere and used as an address for all particles in the universe (the way coordinates do). In order for each particle to have an address, we must continue to use, in parallel, coordinate time. In short, proper time does not mark out all of space; it is a kind of intrinsic parameter of each particle. Its name is most appropriate: "proper" in the sense of "belonging especially"; a personal time.

The reactions to Minkowski's article were far from being all favorable. Jakob Laub, for instance, in a letter to Einstein,[45] felt "for now, even more skeptical toward Minkowski's paper." As for Einstein, he was not comfortable with Minkowski's four-dimensional formalism, and his first references to Minkowski's papers were extremely cautious.

NOTES

1. Einstein, 1921b, p. 123 or 124; English translation in Einstein, 1954, p. 233.

2. Einstein, 1905a, in Lorentz et al., 1923, p. 37.

3. Ibid.

4. On this topic see Lévy-Leblond, 1976 and 1977. On the history of special relativity see Miller, 1981.

5. William Whiston, 1707, quoted by Michael A. Hoskin, 1978, in Taton, 1978, pp. 235–236.

6. William Herschel, an English astronomer who, at the end of the eighteenth century, revolutionized the view of the universe held at the time.

7. William Herschel, 1802, in Dreyer, 1912, vol. 2, p. 213. Quoted in Harrison, 1987, pp. 141–142.

8. William Herschel, 1813, quoted by Michael A. Hoskin, 1978, in Taton, 1978, p. 235.

9. Newton, 1729, p. 8.

10. There is much to be said regarding Einstein's influences on these foundational questions. One such analysis may be found in "Editorial Note: Einstein on the Theory of Relativity", in Howard and Stachel, 1989, pp. 253–274.

11. Regarding the prehistory of special relativity and Einstein's influences, in particular Poincaré, see Miller, 1981; Holton, 1973; Stachel, 1995, and Darrigol, 1995.

12. Poincaré, 1902. In 1952, in a letter to Besso, Einstein remembered these readings, and most notably these two books (Einstein to Besso, 6 March 1952, in Speziali, 1972, p. 272).

13. Einstein to M. Solovine, in Einstein, 1956, p. vi.

14. Ibid., p. viii.

15. Poincaré, 1891.

16. Poincaré, 1898.

17. Poincaré, 1900, in Bosscha ed., 1900. In this regard, cf. *CPAE*, vol. 2, p. 308, n. 10.

18. Poincaré, 1902, pp. 111–112.

19. Whittaker, 1953, pp. 27–77.

20. Particularly by Holton (1973), whose views are clearly stated on page 176: "Since Whittaker's analysis has been and is likely to continue to be given considerable weight, it is necessary to examine it closely. It turns out to be an excellent example of the proposition that no such analysis can be considered meaningful except insofar as it deals both with the material it purports to cover and with the prior commitments and prejudices of the scholar himself."

21. Poincaré, 1905.

22. Poincaré, 1906.

23. Poincaré, 1905, p. 1504.

24. Ibid., p. 1505. He made this precise in his 1906 article (Poincaré, 1906, pp. 132–136). Einstein made the same remark in his 1905 paper: Einstein, 1905, *CPAE*, vol. 2, p. 292.

25. Poincaré, 1906, p. 167.

26. Ibid., pp. 168–169.

27. In this regard, cf. Walter, 1999, pp. 12–14.

28. Holton, 1981, p. 149.

29. Einstein, 1905, *CPAE*, vol. 2, pp. 149–169. The light quanta (or "photons") hypothesis.

30. Einstein, 1905, *CPAE*, vol. 2, pp. 185–222. Einstein's dissertation, submitted to the Faculty of Philosophy of the University of Zurich.

31. Einstein, 1905, *CPAE*, vol. 2, pp. 223–236. This is his paper (very important also) on Brownian motion.

32. Einstein, 1905, *CPAE*, vol. 2, pp. 275–310. This is, of course, his paper "On the Electrodynamics of Moving Bodies," a sketch of special relativity.

33. Einstein to C. Habicht, May 1905. Einstein, 1905, *CPAE*, vol. 5, pp. 31–32.

34. I shall often use the terms "proper," "intrinsic," and "invariant" in the sense of "absolute," or "independent of any particular measurement." It is the exact opposite of "relative" and an essential concept in . . . relativity. Relativity arises from measurements made in a necessarily particular system moving in a particular way and on which one projects a concept that was defined in an intrinsic fashion, the proper physical magnitude, in order to calculate the value predicted by the theory. This may help to understand why, for a while, Einstein called his general relativity theory "the theory of invariants."

35. Einstein to Erika Oppenheimer, 1932, quoted in English translation in the editorial note "Einstein on the Theory of Relativity," *CPAE*, vol. 2, pp. 252–274, p. 261.

36. It is perhaps worth recalling that an event consists of a place and a date, a point or an address in space-*time*: three spatial and one temporal coordinates.

37. Einstein, 1907, *CPAE*, vol. 2, pp. 433–488.

38. Minkowski, 1908, in Lorentz et al., 1923, pp. 73–91.

39. Ibid., p. 75.

40. Ibid., p. 83.

41. Ibid., p. 76.

42. Ibid.

43. Ibid., p. 81.

44. Let us observe, in passing, the extent to which relativity is really the theory of invariance.

45. J. Laub to Einstein, 18 May 1908, *CPAE*, vol. 5, p. 120.

CHAPTER THREE

Toward a New Theory of Gravitation

BEFORE GETTING DOWN to the details of general relativity and analyzing each of the theory's guiding principles, let us try to understand toward which new horizons Einstein is going to take us. What is general relativity? It is a theory of gravitation that replaces Newton's and which predicts the behavior of material and light particles subject to a gravitational field. General relativity disregards all other physical phenomena, such as quantum and electromagnetic ones, to focus only on a classical (i.e., nonquantum) physics, a kind of particle ballistics.

A gravitational *field* is simply a concept, a theoretical tool that expresses the existence of gravitation at each point of space-time, a field that we will be able to construct thanks to general relativity. Einstein's theory of gravitation first tells us the nature of that field, in which space it exists, what its equations are and how to write them, and finally how material and light particles behave in it.

To better understand the changes (the numerous changes, in fact) that will take place, let us go back for a moment to Newton's theory of gravitation. As a consequence of its absolute space and absolute time, his theory operates in an absolute space-time that we shall call *Newtonian*. The position of each particle, material or luminous, is defined by its four coordinates, three spatial and one temporal; they determine where and when. Given a distribution of matter of mass M, the Sun, for instance, we know how to calculate the gravitational field created by this distribution everywhere in the universe. It is essentially an M/r^2 force field. Any test particle, that is, a particle small enough so that we may ignore its own gravitational field, of inertial mass m, will be subject to that force field. But this particle is also affected by its own inertia, a force that is proportional to its acceleration a. It follows from the fundamental law of dynamics that those two forces cancel each other out, and hence we can derive the trajectory of the particle:

$$f = m\,MG/r^2, \quad f = -ma, \quad \text{hence, } a = -MG/r^2.$$

The acceleration of a particle is proportional to the (gravitational—more on this later) mass M that creates the gravitational field and inversely proportional to the square of its distance from it. But this acceleration is totally independent of its (inertial) mass m. Therein lies the equivalence principle of Newton's theory of gravitation. If we know the acceleration, it is easy to obtain the velocity of the particle provided we also know the initial conditions: the position and velocity of the particle at a given moment. Its trajectory is then completely determined: it is the path the particle will follow in absolute space as a function of absolute time.

There is no such determination in general relativity, where the concept of force plays no role, and the gravitational field created by the distribution of matter in Newton's theory is now represented by the curvature of space-time. For this to be possible, the space-time in question has to be "curved." Now, the only space-time we have so far considered—that is, Minkowski's—is flat; it has zero curvature, which is consistent with the fact that it does not represent any gravitational phenomenon. General relativity will therefore involve a new type of space-time that is curved. These curved space-times, known as Riemannian because they were invented by Bernhard Riemann,[1] are to Minkowski space-time as a curved surface is to a plane (if we restrict ourselves to two dimensions, i.e., to planes and surfaces).

Hence, for any given gravitational problem there is a corresponding space-time whose curvature will represent gravitation at each point of it. It is not advisable to try to picture the curvature of such space-time. In relativity, representations are misleading and it is often better to avoid them and be aware of the fact that mental images are ambiguous and should not be given too precise a meaning.

But before we can construct the space-time that is the solution of the given problem, we need equations. The equations of the gravitational field have two sides, as all equations do:

$$E_{\mu\nu} \equiv R_{\mu\nu} - 1/2Rg_{\mu\nu} = \chi T_{\mu\nu}.$$

The left-hand side, $E_{\mu\nu}$, defines the geometry, the structure of the solution space-time, and it will represent, through its curvature, the gravitational field that results from the distribution of matter, $T_{\mu\nu}$ given a priori on the right-hand side. The tensor form of the field equations is deceivingly simple; in fact, it hides ten formidable nonlinear equations.[2]

The fundamental equations of general relativity—the field equations—are a sort of machine to define the space-time whose curvature represents gravitation. In this complex mathematical mechanism, the right-hand side of the equation (also called the matter tensor) represents the distribution of matter, while the left-hand side represents the geometric structure of space by means of variables known as gravitation potentials. The latter roughly express the intensity of the gravitational field from which the curvature of space will result. The series of rather complex equations so obtained must now be solved, and the solution is a curved space-time. The equations of general relativity operate on a collection of space-times; they must decide which one, among the available space-times, represents the geometry corresponding to the distribution of matter that was imposed at the outset by the right-hand side of the equation.

The solution space-time corresponding to a total absence of matter (the empty space, or $T_{\mu\nu} = 0$) is, as would be expected, Minkowski space, which is flat and gravitation-free. However, the equations of general relativity may also give other answers to the question of the geometric structure of the void. This raises certain problems that we will discuss later.

The tensorial character of the equations (expressed by the subscripts μ and ν, which range over the coordinates) should not scare us; it simply means that the equations are written a priori in a completely general fashion, valid in all coordinate systems.

In any particular situation, the coordinate system will be chosen taking into consideration the symmetry of the problem. For example, rectangular coordinates will be preferred if the distribution of matter exhibits a planar symmetry (flat layers of matter), but spherical coordinates are more convenient when working with the Sun's gravitational field. Before we can find the *right* space-time, the one corresponding to the given distribution of matter, we need to specify the limit conditions. In particular, we must choose the structure of the desired space-time at infinity; most of the time it is a flat, Minkowskian structure that is assumed. If we are dealing with a dynamic problem that evolves with time, we must also specify the initial conditions of our *model*.

We are really building a model of space-time to represent the distribution of matter we have imposed on the right-hand side of the equation. It may appear a bit odd that we should build different models of the universe. After all, isn't there only one universe? To be sure, if the mathematical problem were simple, we would solve it once and for all by specifying on the right-hand side the actual distribution of matter for the real universe, and we would derive the true structure of the universe, our space-time. That is precisely the goal of relativistic cosmology, which starts with very rough estimates of the distribution of matter and, paradoxically, obtains rather precise results. (We will come back to this topic in chapter 15.) In fact, the mathematical problem is much too complex, and we must tone down our ambition and tackle only simplified or limited versions of our universe. We must begin by understanding the solutions that general relativity offers for simple but already difficult questions, such as the structure of the space-time generated by a sphere of matter or by a star.

It is difficult, if not impossible, to correctly conceive the curvature of a four-dimensional space-time. We must accept the fact that all we have is a mathematical definition (curvature may be defined in terms of tensors) and content ourselves with some simple images which, even if imperfect, can help us think. The same applies to the curvature of a surface (compared to the plane which has zero curvature); or to that of the trajectories of material particles, the paths of planets and stars; or better still, to the curvature of a beam of light. But these are merely images and not what curvature of space-time precisely means. There is no salvation outside equations! Even more than other theories, general relativity works like a kind of black box from which results in the form of space-times come out; results that we struggle to understand, to interpret, to visualize.

Unlike what happens in Newton's theory,[3] in general relativity, light particles are affected by gravitation; they do not generally follow linear paths but travel along curved trajectories in space-time. This curvature of light beams may be detected by measuring it in some very special astronomical conditions—solar eclipses—and the correctness of general relativity may be verified in this way (see chapter 8).

As for time, it is not the same throughout the whole space-time; it varies along with the gravitational field. If the latter is particularly strong, clocks will stop. But which clocks? What are the clocks of space-time? They are simply the atoms, whose vibrations mark the proper time, *their* proper time. Thus, the frequency of a certain emission line should not be the same on the surface of a very massive star (such as a white dwarf) as it is on Earth, and this frequency will be measured to make sure that general relativity does not lie (see chapter 9).

What about the trajectory of the planets, which we know obey the laws introduced by Kepler and justified later by Newton? What happens when we apply Einstein's theory of gravitation at that level? Not much, really, for our solar system is a very poor laboratory in which to test gravitation, due to the Sun's extremely weak gravitational field. And yet, since the middle of the nineteenth century it had been observed that Mercury exhibits an anomaly that the Newtonian framework could not explain. We shall see in chapter 7 that its resolution, back in 1915, was precisely the first success of general relativity. Why Mercury and not Jupiter, which is infinitely heavier? Simply because Mercury is the planet closest to the Sun, and it is therefore the one subjected to the strongest gravitational pull. In chapters 8 and 9 we shall see that the weakness of the solar gravitational field was not the only problem that general relativity faced in its search for occasions to be put to the test and gain respectability. Almost fifty years passed before physicists (other than the relativists, who make a living with it!) would believe in Einstein's theory and take it seriously, a point that I will elaborate on in chapters 10 and 11. The situation is different today, thanks to the spectacular development of a field known as relativistic astrophysics, which uses general relativity in particular to understand the dynamics of stars and to predict *exotic* phenomena—exotic with respect to Newton's theory, of course.

Before all that, in the 1960s, the relativists would at last turn their attention to the strong gravitational fields and wonder, for instance, what happens if gravitation is so strong that it not only deflects the beams of light but bends them so much that they are twisted back to the star that emitted them. Black holes, which Einstein himself never wanted to envisage, were born in this way (chapters 12, 13, and 14).

Finally, if curved space-time is the child of general relativity, what about the universe? Gravitation is the only fundamental force in physics that acts at great distances, and so it governs the trajectories of stars and galaxy clusters; gravitation rules the universe. Back in 1917, cosmology appeared to Einstein as a fundamental application of his theory—relativistic cosmology, naturally—an application that never ceased to be successful throughout the field's eighty years of existence (chapter 15).

However, to achieve that necessitated a lot of work and plenty of papers, notably Einstein's. He first had to invent, to formulate, a theory of gravitation for the twentieth century. There was nothing more interesting and fascinating than observing Einstein at work. How did he work? How did he navigate through this sea of facts and concepts that had to be made into a logical and simple whole that he also wanted to be harmonious? How did he make the choices that took him to the theory we today call general relativity and to its notorious successes?

CHAPTER 3

RELATIVITY OR THEORY OF INVARIANTS?

Relativity—an ill-chosen term: not only because there are three kinds of relativity, but also because *relativity* suggests "relative" and the idea of philosophical relativism, a position that is almost at the extreme opposite of the spirit of relativity. According to Lalande's *Vocabulaire technique et critique de la philososphie* (*Technical and Critical Vocabulary of Philosophy*),[4] the relativity of knowledge means that we "cannot know things but only relations," and so we can only know "states of consciousness" and not "things in themselves."

Now, relativity (whether special or general) is concerned only with the description of phenomena, a description that depends on the observer and the way the observer moves with respect to the object being described. As for the phenomenon itself (for example, the frequency of a certain atom), it is perfectly determined, known, and well defined in its own frame of reference (which for this reason is called *proper frame*); it is the phenomenon "in itself." The observer in motion will provide a description *relative* to his or her own frame and so project the phenomenon onto this frame, as any other observer would do. Thus, the description of the given phenomenon will depend on the observer, and hence it will be *relative* to the observer, while the phenomenon itself will be *intrinsically* described by a magnitude, an *invariant* object which, as it is projected onto the frames of the various observers, will take different values relative to the trajectories of these observers.

Going back to our example, the proper frequency of an atom will be defined (and measured) in its (proper) frame. But, seen from elsewhere, its value will be a function of the state of motion (and of gravitation) of the place where the measurement is taken.

And so the phenomenon *in itself* will be defined by an *invariant* (which is only a measurable component of reality); in order to be measured, this invariant will be projected (thanks to the theory of tensors) onto the observer's frame, whose motion will affect the invariant's value. It is the description of the phenomenon, and not the phenomenon in itself, that is relative to the observer. The deep meaning of Einstein's relativity derives from the concept of the intrinsic magnitude, of the invariant—a far cry from philosophical relativism.

Einstein did not choose the term *relativity* and would have preferred to call his theory of gravitation "the theory of invariants."[5] Those same invariants were considered by Max Planck to be of fundamental importance in special relativity. He contended that each relative magnitude was necessarily linked to an invariant. To him, the rejection of the absolute nature of space and of time did not do away with a notion of an absolute that he located in space and time blended together in a single continuum: space-time.

When Einstein decided to tackle gravitation, few of his colleagues supported him. What's the point in attacking gravitation, they wondered. Isn't Newton's theory well established? Doesn't it quite satisfactorily apply, not only to our solar system but also to double-star systems? Isn't it, as far as we know, valid throughout the universe?

And yet, had they been consulted earlier, the best theoreticians of the time would have admitted that there was indeed an inconsistency between gravitation and relativity. But they would have tacitly agreed that the question could wait, for there were more pressing things to do in physics, such as working on X-rays, the structure of the atom, or even the uranic rays discovered by Becquerel.

There was, however, something rotten in Newton's kingdom of universal gravitation. As Poincaré had observed in 1905, Newton's theory, based as it was on Galilean kinematics, was simply incompatible with the new kinematics of Einstein and Lorentz. In particular, c, the speed of light, did not appear in Newton's fundamental equations, and gravitation propagated with infinite speed, something that was intellectually unacceptable. This question of the instantaneous action of gravitation was not new, and Newton's theory was strongly criticized in that respect. A new theory of the gravitational field was thus needed in which the speed of propagation was finite, just as in that of the electromagnetic field.

Unlike Maxwell's electromagnetism, Newton's theory of gravitation was logically incompatible with special relativity. It was an inconsistency that Einstein could not stand, and he would guarantee the compatibility of his theory of gravitation with special relativity by requiring that its equations be "Lorentz invariant." There was no choice but to render gravitation relativistic; the consistency of physics was at stake.

There was another reason for solving the problem of gravitation. Through the equivalence of mass and energy—that is, the equation $E = mc^2$ that Einstein had just discovered[6]—special relativity demanded that gravitation be taken into account. If mass is energy, then energy has a mass. The inertial mass (roughly speaking, the resistance that a body opposes to a change in its motion) is involved in this equation; but insofar as Einstein stipulated as a principle that the gravitational mass (basically, the gravity inherent in any mass) be equivalent to the inertial mass, then energy is a source of gravitation. Hence, one is entitled to ask what the consequences of this equation (and therefore of special relativity, from which it was derived) are for gravitation.

That was a question that could not fail to arouse Einstein's interest and excitement. On the one hand, he had to find a generalization of Newton's theory of gravitation to the new relativistic kinematics, and on the other, he had to ensure the validity of this generalization in completely general coordinate systems: two different paths for one and the same project. In 1907, Einstein began quietly working on his project. He was still only a technical officer at the Swiss Patent Office in Bern, but his name began to be known to the best experts in theoretical physics of the time.

THE RULES OF THE GAME

For a theory in physics to work, it must be supported by a mathematical apparatus rooted in the observed reality; a mechanism made up of equations, variables of different kinds, concepts, and a kinematics; in short, a theoretical framework in which the equations become meaningful. And all that—the framework, concepts, observable physical magnitudes, and equations—has to be invented. But how does one go about building and writing a new theory?

The problem is that no one knows how to construct a physical theory because there are no rules. Everybody has his or her own ideas, sensitivity, and ingenuity; his or her own genius. There is no right method or definite rules for writing a literary masterpiece; why should it be any different in the case of a physical theory? Each person has his or her own rules, which evolve along the way, and those used by Einstein to write his special relativity had to be rethought, discussed, and modified before he could construct his general theory. Nevertheless, in Einstein's method, in the way he worked, in his style, there were certain constant traits that we shall try to portray in order to better understand his theories.

To build a physical theory one needs first of all a working space, some raw materials (experiments and laws), some concepts to manipulate (mass, velocity, distance, time, and so on), and also a framework in which to think. A theoretician works with some key ideas which support the whole process; in the same fashion, Einstein constructed his theories with principles.

Like a child who builds a toy house with blocks of various colors, Einstein started with sets of principles, the conceptual blocks or theoretical elements that he could place, move, suppress, and arrange in different ways; these are the bricks he used to erect his theoretical buildings. An architect was at work here, and his basic materials were his principles. He chose those principles with care, even if not all of them had the same strength, the same soundness, the same degree of truth. Within this theoretical framework he had a certain freedom, an essential condition for setting up a working space that would only gradually, from 1907 to 1915 (see chapter 5), take the form of a consistent theory.

In this theoretical construction game there was a priori a huge number of degrees of freedom. Many relativistic theories of gravitation were conceivable, and Einstein swam in this sea of possible theories. But which one was the right one? Which one would satisfy his principles? Which one would account for the observed facts? Was there a right theory at all? The problem was so difficult to translate into equations that it was essential to establish a working framework, and that was the purpose of the set of principles. He would work, think, dream, and in a sense have fun with those principles. There is no question that during those years of transgressions in search of the right theory of gravitation, Einstein had a good time, which did not prevent him from occasionally having doubts about the path he had chosen. But it is primarily with the principles that he played, that he worked; he used them a great deal and remained serenely faithful to them.

If what we call genius does exist, then it was certainly at work in Einstein's case. If it were known how to invent a physical theory, we would not be talking so much about the subject. During this stage of creation, it is essentially genius that it takes; it is, in a strong sense, an art: the art of choosing a canvas, a framework that nature will accept, of accounting for reality. It is the art of selecting the right angle, the perfect point of view, and of ignoring certain features of reality which the artist a priori believes, rightly or wrongly, to be accessory. The artist must know how to choose correctly so as not to get lost in the sea of facts, and still allow a hitherto unknown wind or undercurrent carry her to a new shore, to an unexplored land where a new world can be built. Not only must the artist know to emphasize a certain point or choose a certain perspective, but she must also firmly believe in it and hold the course despite the doubtful frown, the disdainful look, and the sarcasms of certain colleagues.

It is an art where the artist has the choice of subject, medium, and tools. The work will be judged on its merits, on its results, on the way experience and observations fit its propositions. The witness of this art is not, as in painting, the public visiting the galleries; its "supreme judge" is experience, as Einstein would forcefully write.

During this period of creation, Einstein had to put together his principles in the context of the many possible structures: to begin with, spatial structures, that is, the different types of geometry; and, within each type of space, he tried out several equation forms compatible with both his principles and the given structure.

The architect of the universe was at work making plans and drawing outlines. "Outline" is precisely one of the words in the title he chose for one of his first theoretical essays on gravitation. And if we assumed . . . ? And if I . . . ? What would happen if . . . ? Will that work? In the silence of his study he was inventing the universe, a representation of the universe.

But what do we precisely mean by "principle"? This term represents, not only for Albert Einstein, different realities. It is a word that appears quite often in his writings, that he employed in discussions with his colleagues, and which he used, above all, to create his theories: the principle of thermodynamics, the principle of the constancy of the speed of light, the principle of relativity, the principle of covariance, the principle of equivalence, Mach's principle.

Let us try to be more precise about what we should understand by "principle." It is a word that suggests a relatively static vision of the construction of a theory; it is stated at the beginning of the incipient theory and refers to its foundations. A principle is a kind of primal reason, but it may also be a law of nature that the theoretical physicist puts forward as one of the bases of the theory. That type of principle must be understood in the context of reality as a sort of rule that nature *appears* to obey, a rule with no exceptions. It can also be a requirement stated at the outset as a postulate, something in which one believes and that is not to be questioned: thus, the *Principia*, which form the foundations of Newton's natural philosophy, foundations on which he built his theory and which were validated a posteriori as a result of the interest and the power of their consequences.

Those two points of view, those two levels—physical structure and mathematical reality—are found in the way physicists, and in particular Einstein, attempt to sort out the universe. Theoretical physics combines in an inextricable manner the two levels, the natural and the mathematical, the real and the formal ones, and even more so general relativity, which transformed geometry, until then regarded essentially as a branch of mathematics, into a branch of physics (see chapter 6).

In another context, a principle is a rule of conduct, a kind of standard that one imposes on oneself or on others, as in the expression "to have principles." This moral connotation is not innocent; to have principles, to stand by one's principles, suggests a rather strict condition from which one does not depart easily. But to what extent was Einstein strict regarding his principles? Reality and experience are obviously stronger than principles, a truth that applies particularly to scientists, whose task is first and foremost to account for natural phenomena.

Einstein's principles were often connected to, or based on, experience, namely, the principle of the constancy of the speed of light, the principle of equivalence or the principle of relativity. But the observed regularity of a law can only be verified up to a certain degree of accuracy; there is always a margin between the experimental law that is necessarily approximate, marred by the uncertainties of measure, and its formulation as a principle that cannot tolerate any exception. To state an experimental principle amounts to formulating a (small) theory, and it therefore involves a serious risk of refutation. If such is the case, if some experiment is at variance with the principle, it will bring the whole edifice down; the principle will then have to be abandoned and the building started again on new foundations.

When Einstein contemplated—he actually did!—calling into question the principle of the constancy of the speed of light, it was within the limits of precision of experimental results. The theoretician has the freedom to make such decision, a freedom that only experience may, one day, deny him or her.

All these principles and others, which are either a kind of minitheory or the expression of some philosophical—or partly physical, partly philosophical—choice, helped Einstein to have in hand and articulate some very profound ideas and a variety of experiences. Principles were the basic elements of his work; they allowed him to see things with a certain detachment and under a simpler light.

But in order to construct a theory, one must respect a number of conditions and known facts. If the bases of kinematics were to be reformulated, the new theory—special relativity—had to explain *all* that Galilean kinematics explained *and* also be compatible with Maxwell's equations and consistent with the experiments on light carried out by Arago, Fizeau, and Michelson and Morley. In the case of a new theory of gravitation, it had to explain *all* that Newton's theory of universal gravitation enabled us to understand, beside, in the first place, accounting for Kepler's laws.

Also, the new theory had to be consistent with special relativity; otherwise what would be the point in modifying Newton's theory? And finally, it had to account for certain minute discrepancies in the trajectory of the planets that Newton's theory could not explain. But such was not its *first* priority—a source of criticism to which we shall come back. The preceding incomplete enumeration should be enough to realize that it was nearly impossible to hold in one's hand all those elements at the same time, not to mention manipulate them. The principles were thus the cards of a theoretical game, each representing some feature of reality.

Looking for principles amounts to searching for the rules of the game—the game of the universe. Einstein was no doubt tempted to see his principles in that way. To what extent did he really believe that analogy? Had he ever thought that his principles were anything more than the blocks of a theoretical construction game, that they expressed some really universal Law? Did he believe so? There are of course different ways of believing, various degrees of belief, and many meanings of the term. There is "belief," and then there is "Belief." Sometimes, there is hope in the belief. Einstein laid down principles hoping that they would open for him the door he had in mind, especially since principles and their consequences have different meanings depending on the context and the way they are organized. The principles were the themes with which he composed the symphony; some were discordant, others did not conform to facts or lead to absurdities but sometimes they converged toward the same goal.

CHAPTER 3

EINSTEIN'S STYLE

After having completed special relativity in 1905, Einstein was in his heart convinced that he was on the right track, and perhaps he even thought to have found, if not really a method, at least a style that would allow him to move forward. His certainty stemmed from his successful analysis of physics, from the manner in which he had obtained his results, and, foremost, from his principles. He thus set out to work on general relativity encouraged by the success of his style in the elaboration of special relativity.

A theory built on principles will exhibit an admirable and severe architecture, but it will also be highly rigid and hence not without a certain brittleness. Although principles are obviously very restrictive, it is precisely in this a priori restriction of the theoretical context that their appeal lies. The analogy with a construction game applies here as well. The moment we lay a particular set of principles on the table, the question is, if not solved—for an infinite number of theories may conform to a given collection of principles—at least circumscribed.

But Einstein's attitude was a complex one and depended on the context. He was extremely demanding but not dogmatic and, even if he found it difficult and at times painful, he would change his mind and try to adapt himself to the situation. And so, much as he would have wished, he was not able to hold on to all his principles and often had to yield to observation and to the theoretical reality. But as he built his theory of gravitation he would not give up the principle of relativity, simply because (special) relativity held water—the water of facts. It was at this level that Einstein performed as a virtuoso and showed an extraordinary physical sensitivity and a great freedom of judgement.

Beyond the principles, Einstein insisted on the theoretician's freedom, a term that he frequently employed in his epistemological writings. It was precisely in the context of theoretical constructions that he defended this freedom. Far from letting his principles stand in his way, he used them as a framework for his thought; an indispensable framework, given the almost unlimited freedom enjoyed by the theoretician at work: the space of possible theories was immense. The moment of discovery, or rather of invention, was the moment of freedom. Surprisingly, it was also the freedom to lay down principles, principles with which he knew how to make compromises. One gives and one takes but one moves forward. Didn't he write at the end of his life that "[the scientist] must appear to the systematic epistemologist as a type of unscrupulous opportunist"?[7] He also said, with his proverbial enthusiasm, that the principles behind special relativity (the principle of relativity in conjunction with the principle of the constancy of the speed of light) "are not to be considered as a 'closed system,' in fact not as a system at all, but merely as a heuristic principle."[8] He certainly did not wish to be fenced in by principles, not even his own, and much less be locked inside a system. He used principles in an essentially *heuristic* way: to make progress in his research or, more precisely, to have in hand the key elements of his work, and to think.

He thus created a style for doing physics that gave a very peculiar and logically tight structure to his relativity theories; one that fascinated theoreticians fond of formal systems and seduced more than one philosopher. While some physicists, in particular those who called themselves relativists, were convinced, others, notably (but not exclusively) the experimental physicists, were not able to make head or tail of it all and lost patience.

Actually, whenever a concept appeared to him useless, ill-defined, or cumbersome, he wondered whether it was really necessary. If it wasn't, why keep it? He had given himself the freedom and needed room for rebuilding the world; he could not keep all those old and useless pieces of furniture. Was Einstein a minimalist? It may seem a strange, daring point of view, but it was nevertheless an important aspect of his manner of doing things. Working with a minimal number of concepts was a constant feature of his physics. Also, if a given magnitude or concept was not well defined, he was tempted to do without;

that was another part of Einstein's style. He turned the problem on its head; he saw it in a different light. And this was, he wrote, "precisely what the theory of relativity has done in a systematic way."[9]

Einstein realized that many notions that appeared to be essential in the classical theories were not at all necessary and were even a burden; they were useless, outdated, and meaningless. Not only were the concept of ether and the notion of absolute space more than debatable, but that of preferential motion was questionable, and absolute simultaneity was only a myth, as he had already shown. He made frequent use of this minimalist approach and even of "an epistemological point of view," stating a sort of ad minima principle: "There is no notion in physics whose use is a priori necessary or justified."[10] And so, in relativity, the notions of absolute simultaneity, absolute speed, and absolute acceleration were rejected because no unequivocal connection with experience seemed possible. This would also be the case, as we shall later see, of the notions of plane, straight line, distance, and so forth, on which Euclidean geometry was built.

One may be amazed to see thrown away so many physical magnitudes, so many concepts that were considered absolutely indispensable to understand space and to do physics. That is precisely the key element in the acceptance of this relativistic revolution that demands of each one of us that we perform a heavy-duty conceptual cleaning job, a true revolution in our way of thinking space-(time)—a recurrent task, so deeply rooted in us are those habits of thought. On arriving at some unknown city, don't we willingly accept getting lost the better to be immersed in it?

General relativity introduced surprisingly few operational physical concepts, and a large part of the scientific literature set out to define new ones. Clearly, one had to be ready to forgo absolute space and absolute speed. It was already a bit more difficult, even problematic, to render relative the notion of acceleration. But it became really strange to have to renounce so fundamental a notion as distance. It was nevertheless the lesser of two evils and even an absolute necessity.

NOTES

1. In fact, the spaces used here are slightly more general than those invented by Bernhard Riemann. They are "pseudo-Riemann" spaces, which are to Riemann spaces what the Minkowski space is to Euclidean space. In those "pseudo-Riemann" or "pseudo-Euclidean" spaces, there is no longer a "trivial" distance. A more general distance is introduced, namely, proper time. In Euclidean geometry, if the distance between two points is zero, they coincide. Such is not the case here: if the square of the distance between two events is zero, these events are not necessarily identical.

2. $E_{\mu\nu}$ is therefore the Einstein tensor and $R_{\mu\nu}$ the Ricci tensor, while R is the scalar Riemannian curvature. $T_{\mu\nu}$ is the matter tensor and χ is proportional to G, the gravitational constant that appears in Newton's theory.

3. We shall see that there are two interpretations of Newton's theory regarding the effect of gravitation on light rays. The simplest one amounts to assuming that light particles are not affected by gravitation: these particles, now called photons, travel in a straight line in a gravitational field. This is the most common explanation and the one we shall adopt here. But one may also assume, as the natural philosophers of the end of the eighteenth century did, that since a light particle has a mass (in the context of Newton's corpuscular theory), it may be subject to gravitation just as any other material particle. In a strong gravitational field such as it exists near a star, its trajectory would be curved, deflected.

4. Lalande, 1960, p. 914.

5. It was the mathematician Felix Klein who suggested in 1910 the name "Invariantentheorie." See "Editorial Note: Einstein on the Theory of Relativity," in *CPAE*, vol. 2, pp. 253–274. *Cf.* also Holton, in Holton and Elkana, 1982, p. xv.

6. Einstein 1905b, in Lorentz et al., 1923, pp. 67–71.

7. Einstein, 1949, in Schilpp, 1949, p. 684.

8. Einstein, 1907, German original in *CPAE*, vol. 2, p. 410; English translation in *CPAE*, vol. 2, p. 236.

9. In an annex to Einstein's letter to Solovine, 24 April 1920, in Einstein, 1956, pp. 19–21.

10. Ibid.

CHAPTER FOUR

Einstein's Principles

According to general relativity, the concept of space detached from any physical content does not exist.[1]

WE HAVE SEEN the reasons that convinced Einstein to turn his attention to gravitation. But where was he to start? How was he to proceed? It was an amazing idea that prompted him to take the plunge and get down to work. It was, in his own words, "the happiest idea of my life."[2]

Strangely enough, it concerned a very ancient experiment, the one Galileo is supposed to have performed at Pisa (but which was probably older and already by then quite ordinary) and which suggested that all bodies fall in the same way regardless of their mass, composition, or nature if air resistance is not taken into account. Once again, Einstein reversed the problem and saw it *differently*, from *another* angle—an angle which, once we understand it, once we *see* it, appears so obvious, so trivial. But that reversal, that step Einstein took, was one that nobody until then had dared to take or realize its implications. What Einstein did is called thinking.

Starting then from this seemingly trivial experiment, Einstein realized that he would be able to take a leap forward, and what a fundamental leap it was: he would be able to treat gravitation within the framework of special relativity. From then on, he would be fascinated and obsessed by this idea which enabled him to start laying the foundations of a relativistic theory of gravitation. It immediately became the very first principle of general relativity: the principle of equivalence, the theory's driving force. For Einstein the idea took the form of a daydream, of a man "falling freely from the roof of a house."[3] He believed that for such an observer "there exists—at least in his immediate surroundings—no gravitational field."[4] Everything happens as if he were at rest, and any objects that he may drop during his fall will remain around him in a state of apparent rest or of uniform motion. In short: the gravitational field is absorbed, cancelled. Physically speaking, there was nothing really new in this image, except for a change in the point of view: where Galileo saw only bodies fall, Einstein himself fell or, more precisely, he made his observer fall—out of precaution. This meant that it was possible for a moment to free oneself from gravitation and afterward, on returning to the initial reference frame at the foot of the tower, to try to understand what was going on. This thought-experiment, this idea, provided him with a *state of motion*—free fall—in which he could work *outside gravitation*, and it offered him an amazingly powerful working tool.

GRAVITATIONAL MASS AND INERTIAL MASS

When I was a child, at school, there was a long glass tube standing in a corner of the classroom. One day, the teacher brought it to the front of the class and placed a small piece of cotton and a small piece of lead inside it. The tube was closed at one end with a brass lid; the other end had a nozzle through which it was possible to create a vacuum. I have no idea how the teacher produced the vacuum, but I clearly remember him perched on the platform holding the tube, which he quickly turned over so that the two objects inside could fall. He had previously asked the class which one—the cotton or the lead—we thought would reach the bottom first. "That's too easy, sir," we had replied almost in unison. "The lead, of course!" The tortoise and the hare; we knew the story. The teacher beamed with satisfaction when we all saw that, both objects having left the top of the tube at the same time, the cotton did not reach the other end any later than the lead. By gosh! He then explained that if the leaves from the chestnut trees in the schoolyard took much longer than the chestnuts to fall to the ground, it was simply because the air made them swirl. My goodness! The class was stunned.

This experiment is performed nowadays by every astronaut inside the satellite that is a spacecraft, and we have become accustomed to see on television people and objects swim in space inside—or near—their satellites. The astronaut brushes his teeth, then lets go of his glass, his brush, and his tube of toothpaste, and all those objects remain at rest with respect to him and the satellite. Nothing moves, regardless of the mass or the density of the object—provided the satellite is in orbit, that is, amazingly, if it is falling. It is as if the gravitational field did not exist; it is weightlessness, the proof that one can suppress locally—but only locally—the gravitational field.

Conversely, at liftoff, the satellite is subject to an acceleration of two or three times g, three times the pull of gravity on Earth. But in which gravitational field? For in those $3g$, where is their own g, and where do the other $2g$ come from? In short, is it possible to tell apart the gravitational field due to the Earth from the acceleration field due to the engines? Can we distinguish gravitation from inertia? Would the gravitational field be relative? Wait a second, have we said "relative"? In fact, for Einstein, this *relativity* of acceleration was crucial, and it was the key element that would mark his theory. We shall see to what extent this *total* relativity of gravitation and of inertia was important to him.

We have to go back for a moment to Newton's theory of gravitation. In his law of gravitation, there are in fact two different masses—two concepts of mass are involved: the gravitational mass (the one which *creates* the gravitational field) denoted by m_g, and the inertial mass, m_i (which represents the reaction of a body to forces in general and to the force of gravitation in particular). In electromagnetism, those two concepts are clearly different; an electron possesses a charge and an inertial mass. In a theory of gravitation, the gravitational mass is the equivalent of the charge in electromagnetism. Simply put, the gravitational mass is the gravitational charge.

The inertial mass plays a role in the fundamental equation of dynamics ($f = m_i a$, where a is the acceleration), while the gravitational mass is involved in the definition of the gravitational force exerted by a body of (gravitational) mass M_g upon a body of (gravitational) mass $m_g (f = -Gm_g M_g / r^2)$. And it is because the gravitational mass of a body is—exactly (but why?)—the same as its inertial mass ($m_i = m_g$) that the trajectory of a particle in a gravitational field is independent of its own mass; the acceleration g depends only on the (gravitational) mass M_g that creates the gravitational field and not on the mass of the particle that is subject to it: $g = -GM_g / r^2$. This remark seems almost innocuous. For Newton, it was no doubt a fundamental observation but absolutely not a principle, and we could quite easily assume, without changing anything in the structure of his theory, that the gravitational and the inertial masses are not the same. If such were the case then, in vacuum, a grain of lead would not react in the same way as a grain of wheat (or a small piece of cotton) to a gravitational field. Their orbits would be different for the same initial conditions, as if the grains would be subject to slightly different gravitational fields. That is precisely what happens in electromagnetism, where the ratio e/m is everywhere present; but such is not the case for gravitation.

We must turn to experience if we wish to be convinced. Do two different substances, whose respective physicochemical compositions are as different as possible, react in the same way to a given gravitational field? Well before Galileo, the answer was certainly yes. But Newton wanted to check this by himself. Using wood and gold, he showed that the period of a pendulum was independent of its composition.

At the end of the nineteenth century, Baron Roland von Eötvös carried out some very precise experiments that confirmed the equality of the gravitational and the inertial mass up to a high degree of accuracy.[5] Eötvös employed a torsion balance originally designed to perform local measurements of the Earth's gravitational field, a balance of the same type as the one Henry Cavendish had used in the nineteenth century to measure G, the constant of gravitation.[6]

Two balls of equal mass but different composition were attached to each end of the balance beam, one made of aluminum and the other of platinum. Both masses were subject to the same gravitational field—the Earth's—and also to a centrifugal force in the direction perpendicular to the Earth's axis of rotation. Each ball was thus subject to a force made up of a gravitational component (involving m_g) and an inertial one (due to the centrifugal force, involving m_i). The stronger the centrifugal force, the larger the deviation; but the stronger the gravitational force, the smaller the deviation. Thus, the force acting on each ball would be proportional to the centrifugal force and inversely proportional to the gravitational one—in other words, proportional to the inertial mass (on which the centrifugal force depends) and inversely proportional to the gravitational mass (which governs gravitation). In short, the force applied to the ball would be proportional to m_i/m_g. In this balance the angle of rotation could be measured with extreme precision thanks to a small mirror attached to the axis and on which a beam of light was reflected. If the ratio between the inertial and the gravitational forces depended on the nature of the bodies, the mirror should turn by an angle related to the ratio m_i/m_g. The absence of rotation would mean that the "principle of equivalence," which states the equality of the gravitational and the inertial mass, is true.

Eötvös and his colleagues repeated this experiment several times, in 1889 and at the beginning of the twentieth century, each time with balls of a different material. As expected, the result was always the same: there was absolutely no rotation. This was known as a "zero experiment," from which follows that, up to a high degree of accuracy, the principle of equivalence was verified. More than sixty years later Eötvös's experiments were performed again in various forms and on numerous occasions. In 1961 at Princeton University, Robert Dicke measured the Sun's attraction for bodies of various compositions and showed with remarkable precision (10^{-11}, and even 10^{-12}—that is, with a precision of one part in a trillion—in an identical experiment carried out in Moscow by a Russian team) that those materials behaved in the same fashion with respect to gravitation. This was, and still is, one of the most precise experiments in physics.

Recently, certain ideas derived from string theory seemed to suggest that the equivalence principle and Newton's $1/r^2$ law (which for short distances is extremely close to what general relativity predicts) might not be exact.[7] But some very sophisticated experiments have confirmed the principle of equivalence to 3×10^{-13} and Newton's law to one-tenth of a millimeter. Moreover, thanks to the "moon laser," the principle of equivalence was tested on the Earth-Moon system with comparable accuracy.[8]

THE PRINCIPLE OF EQUIVALENCE

It seems that in 1907 Einstein was not aware of Eötvös's experiments, but that is beside the point, for the real question was whether the equality of the gravitational and the inertial mass had some profound meaning that remained to be discovered: whether it was an accident of the theory, as Newton's equations suggested, or rather a fundamental identity. Whatever the case, the equality of the two kinds of mass was a hypothesis that Einstein made into a principle, and on which he would build his whole theory: the principle of equivalence.

By assuming that the principle of equivalence was a fundamental one Einstein took a considerable risk, for all subsequent developments would then depend on that principle. One day, should some experiment (in the atomic domain, in particular) call into question the principle of equivalence, the entire theory would then collapse and all his efforts would have been in vain. Thus, the very structure of his theory presupposed the validity of that principle. On the other hand, that considerable risk went hand in hand with a no less considerable gain, for this hypothesis would allow him to think about the theory in a different light. Indeed, if the motion of particles subject to a gravitational field was independent of their inertial masses, why should these masses appear in the equations at all? To make them disappear just as quickly, as is done in the Newtonian equations? What's the point of having m in the first two equations above since it is absent from the third one? For the trajectories of test particles in a gravitational field depend only on the field and not on their reaction to the field. In short, to truly believe in the principle of equivalence was to assume that the trajectories are independent of the particles themselves, that the planetary orbits are in some sense traced out in space, in the structure of the gravitational field created by a given distribution of masses. Everything had to be rethought, beginning with the distinction between gravitational and inertial mass, and therefore the fundamental law of dynamics ($f = ma$) had to go, together with the notion of force, which no longer had any meaning. It was quite a program! Everything had to be built from scratch. Nothing was left of Newton's theoretical structure.

The theory to be built would have no use for the concept of inertial mass. The trajectories of particles would be a priori independent of the (inertial) mass of the bodies subject to the gravitational field; they would be directly determined by the (gravitational) mass that creates the field; they would be in a certain sense engraved directly in space, like the paths of rivers on Earth.

Thus, just like Newton's, Einstein's theory of gravitation began with the story of a fall, although this time the protagonist was not the apple but the physicist himself. We shall see that the renewal of Einstein's theory in the 1960s would be symbolically rooted in another, much more spectacular fall: that of an astrophysicist into a black hole. However, it would not be the image of a man falling from a roof, but rather that of an elevator which would popularize Einstein's principle of equivalence.

The person in this elevator is subject to three forces: gravity, of course, and the inertial force transmitted to the cable and the car by the motor, but also the force exerted by the car's floor. For it is the floor that is at the center of everything and which represents, if I may put it that way, the forces acting on our passenger. Without the floor, no force at all, no inertia or gravity: it is the ultimate fall! The passenger falls, flies; it is the free fall like the one inside an orbiting satellite. But the floor, just like our passenger, does not distinguish in this single force the part of gravity from that of inertia: the global force (inertia + gravity) is the weight of the passenger, one that will be greater when the elevator accelerates going up than when it accelerates coming down—and *nonexistent* if the cable breaks. The weight (and hence the force) is relative, that is all; relative not only to the gravitational field (the passenger would weigh less on the moon) but also to the balance that is the (accelerated or decelerated) elevator. We know perfectly well that the astronaut inside the spacecraft being propelled is subject to an acceleration of $2g$, and that his or her weight will then be three times what it is on Earth ($2g$ coming from the acceleration field plus g from the gravitational field); but once the spacecraft becomes a satellite, this weight will be zero.

Hence, locally, there is no *observable* difference between the behavior of a mechanical system in an accelerated reference frame and in an inertial reference frame in which there is a gravitational field. Einstein assumed "the complete physical equivalence of a gravitational field and a corresponding acceleration of the reference system," from which he *inferred* the equality between gravitational and inertial mass.[9] And so it was possible to *locally* substitute a gravitational field with an inertial one, that is, with an accelerated reference frame.

Let us go back for a moment to our satellite—a very special elevator—at liftoff. One can understand, and Einstein himself used this image, that the acceleration felt by our astronauts could not be distinguished from the gravitational field that a (fictitious) planet with a mass two or three times that of the Earth would create. Or, assuming that on their takeoff from the Moon our astronauts are subject to an acceleration of just one g ($1/5g$ coming from the lunar gravity and $4/5g$ from the rocket's engines), they could have the impression, from a gravitational point of view, of being at rest on the ground at Cape Canaveral, where they would also experience $1g$. An acceleration field cannot be distinguished from an ad hoc gravitational field; never, at least during an isolated and local situation.

The above *thought experiment* has a rich conceptual content. Simply by imagining, by thinking, we have been able to understand that acceleration, gravitation, and inertia are concepts of the same kind and can be defined in a way consistent with the principle of relativity. There is something magnificent, even poetical in the clarity of this reasoning process, for we have the impression of getting to the heart of things in an effortless and natural way.

Thus, without any noise or physical apparatus (the experiments were only carried out later), Einstein has allowed us to understand, simply because he saw in a different way what was already known and drew a new (but how simple!) conclusion that no one had glimpsed before. He changed the point of view, at precisely the moment the man begins to fall: he falls and gravitation disappears. It was realized for the first time that, at the instant of the fall, gravitation no longer exists. The fall absorbs gravitation. From one moment to the next, in this thought experiment, gravitation vanishes. This is something strange, extraordinary, perhaps because we are used to identifying the fall with gravitation and we now discover that the opposite is true: it is when we do not fall that we are subject to gravitation's pull due to our weight, which ceases to exist the moment we begin to fall.

But that is not all, because we have not drawn all the consequences of this equivalence. In his initial calculations, Einstein replaced a homogeneous (locally constant) gravitational field by an acceleration field: a sort of sleight of hand that enabled him to find out how gravitation works by knowing only that it obeys the principle of equivalence. From this he drew some *positive* conclusions that we shall discuss in the next chapter. But first, let us go back in time to the very beginning of the century.

MACH'S PRINCIPLE

In his *Mechanics*,[10] which Einstein had carefully read during the meetings of the Olympia Academy, Mach criticized in particular, Newton's concept of absolute space.[11] He denounced Newton for "contradicting his own intention of studying only *facts*."[12] What were the root reasons of Mach's objections to Newton's absolute space? Just as an absolute translation in space is "meaningless" (according to the principle of relativity),[13] we cannot tell an absolute rotation from a relative one. There are in fact only relative motions, translations as well as rotations. Now, in the Newtonian theory, the absolute character of rotation is not only the basis for the calculation of centrifugal forces but also for the

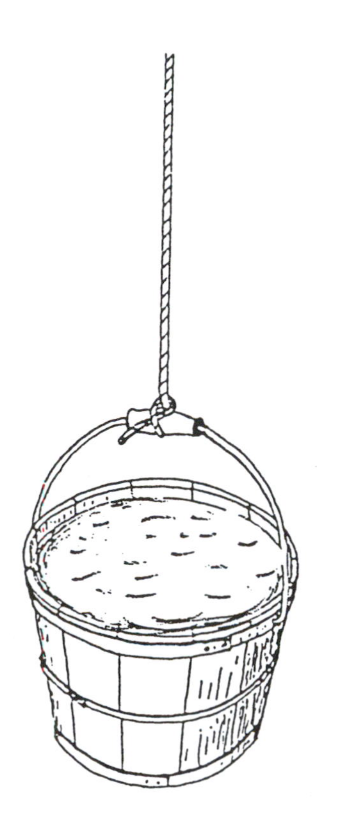

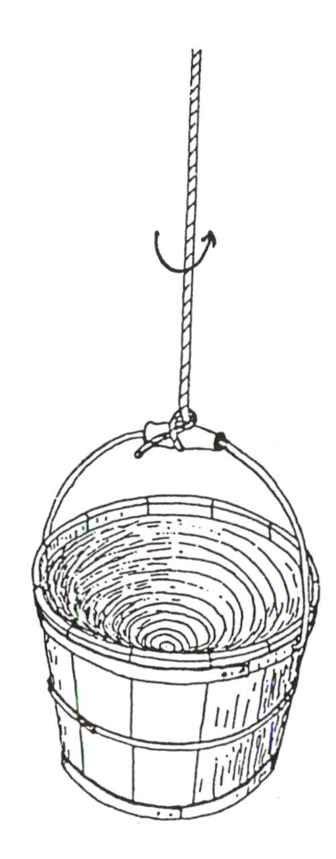

Figure 4.1. Newton's buckets.

reasons in favor of the absolute character of space. There is no doubt that Einstein was convinced by Mach's argument.

Einstein recalled the "two buckets" thought experiment in Newton's *Principia* and Mach's criticism of it. Both buckets are full of water but one of them rotates and the other does not. The surface of the water in the revolving bucket will recede from the middle and rise to the sides of it, but no such thing will happen to the water in the other. For Newtonians, the rising of the surface of the water is a consequence of the centrifugal forces that develop with respect to absolute space. However, absolute space is not a material thing but an intellectual construct, so that this explanation sidesteps the question of the *real* cause of the water rising in the bucket—or of the distortion of a rotating sphere, for that matter. Newton's answer, *absolute space*, was not satisfactory, Einstein said; it is "a merely factitious cause and not a thing that can be observed."[14] Einstein denounced the divorce between the real universe, which is filled with matter, and Newton's immaterial absolute space, and it is undoubtedly this question that must be considered the source of his interest in cosmology (see chapter 15).

This same question manifests itself even more plainly in Foucault's pendulum experiment,[15] and it was taken up by Mach as well as by Einstein. At the poles, the plane of oscillation of a pendulum remains fixed with respect to the fixed stars. Why doesn't this plane rotate along with the Earth? For what reason does the pendulum remain in the same position with respect to the fixed stars? Why does it *choose* to fix the fixed stars? What is the reason for that choice, its material cause? Since the pendulum is attached to a dome that is rotating along with the Earth, why does it choose to follow the motion of Newton's absolute space rather than that of the terrestrial horizon? To hold the pendulum in a plane that is fixed with respect to absolute space (i.e., to the fixed stars) would necessitate a special (inertial) force, together with a *physical* reason for it. But where would this force come from and for what physical reason? It cannot be created by absolute space, and the latter cannot be the support of such a physical connection, for on which mechanism would it be based, then?

More precisely, in the context of Newton's theory, centrifugal forces develop with respect to absolute space as a result of a rotation that is considered to be absolute. In the same (Newtonian) context, if we assume that the fixed stars revolve with respect to absolute space while the Earth or the bucket remain fixed with respect to it (although they are in motion relative to the fixed stars), then the relative rotation of the bucket, the Earth, and the pendulum with respect to the fixed stars would not legitimize the appearance of the centrifugal forces; it is only their rotation with respect to absolute space that can account for those forces. Thus, in Newton's theory, the existence of centrifugal forces due to rotation is a guarantee, a proof of the absolute character of

Figure 4.2. Leon Foucault. (Photograph Paris Observatory)

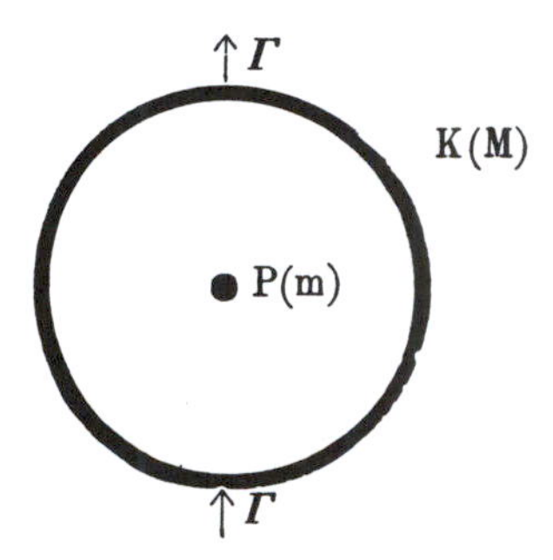

Figure 4.3. The black circle represents the rotating fixed stars whose effect on the small mass at the center Einstein sought to establish. (A. Einstein, 1912, *Collected Papers of Albert Einstein* (*CPAE*), vol. 4, p. 175)

space. But what is this ghost that affects everything? What is absolute space?

Furthermore, if we took away the fixed stars and emptied space of all matter, what could be said? Would rotation with respect to space still generate centrifugal forces? This is an extreme thought experiment to which Einstein was not averse. He gave a negative answer to the question: nothing would happen; there would be neither centrifugal forces nor inertia. For him, centrifugal forces were associated with the action of distant masses, of the fixed stars, and not with the action of a chimerical absolute space.

Around 1912, using for the first time Minkowski's mathematical formulation in the form of a line element of space-time, Einstein built a first relativistic theory of gravitation in which the speed of light c plays a key role, that of supporting gravitational interaction. On the basis of this first theoretical draft, Einstein put his theory through the test of Mach's ideas in a short article published soon after in a little known journal. The sketch that appears on the first page of the article shows a sort of hollow spherical top with a test mass at its center. It is a very schematic model in relation to the question of the (physical) cause of inertia; the top represents the cosmos, the rotation of the fixed stars.

This model allowed Einstein to assess the effect of his hypotheses on the test particle's inertia.

Einstein did not restrict himself to the relativity of rotation but raised, in general terms, the question of the *reason* for inertia; he tried to understand inertia from a physical point of view as a resistance to motion, a resistance that would be connected to the mass of all the other bodies in the cosmos.

It is important to realize that, from a physical perspective, we are in the presence of two completely different phenomena: gravitation and inertia. Gravitation requires the interaction of *two* masses, while inertia manifests itself as soon as a single mass is present: the sling is a good example, for it tightens as it is swung around. But where is the interaction that explains this centrifugal force, where is the second element of this *inter*-action? What is the physical reason for the tension? How does one stop it? And what if we conceive of inertia, of which the centrifugal force is one of its manifestations, as the result of the action of distant masses, of the fixed stars, of the cosmos on the motion of masses down here on Earth? If we could make the cosmos disappear with a stroke of the pen, wouldn't inertia vanish, too, just as gravity vanishes when one of the two masses is suppressed? If the reason for gravitation is the interaction of two masses, inertia only exists by virtue of the action of absolute space—or of its material expression, that is, the cosmos—on each of the particles of nonzero mass that make up the universe. In short, gravitation is in effect an *inter*-action, while inertia is an action of that ghost that is absolute space. However, there is a reason why we so easily confuse gravitation with inertia. This confusion stems from the equivalence between inertial and gravitational mass, from the principle of equivalence. Inertia and gravitation are inseparable; there is no way to tell them apart except intellectually. Then, why not regard them as one and the same phenomenon? Let us abandon the idea of inertia, which would simply be the gravitational action of the cosmos on the totality of all masses.

Einstein wanted to regard inertia and gravitation as a single phenomenon seen from two different points of view. His initial results appeared to him as decisive: "The total inertia of a mass-point is a consequence of the presence of all other masses, which is based on a kind of interaction with the later."[16] This shows the profound effect that Mach's criticism of the Newtonian concepts had on him. Einstein thus added one brick to his edifice, one principle to his conceptual network, an idea he would perhaps value even more than the others.

Once the decisive step was taken, that is, resorting to a Riemannian space, he immediately sent his article to Mach, together with a letter sketching out the essence of his hopes at the time: "You probably received a few days ago my new paper on relativity and gravitation, which is now finally completed after continual exhaustion and tormenting doubts. . . . If so, then . . . your brilliant investigations on the foundations of mechanics will have received a splendid confirmation. For it necessarily follows that the origin of *inertia* is some kind of *interaction* of the bodies, exactly in accordance with your argument about Newton's bucket experiment."[17]

In a last letter to Mach, at the end of 1913, he acknowledged the fragility of those arguments that, as he was well aware, were only of an epistemological nature. It was clear that the idea according to which inertia would be due to "distant masses" could not be experimentally tested. But that did not prevent him from thinking that it was "absurd to ascribe physical properties to 'space.' " . . . The reference system is so to speak, tailored to the existing world," and, he added, "[it] loses its nebulous aprioristic existence."[18] Space was no longer something given but something yet to be constructed.

Einstein devoted a considerable amount of work, energy, and interest to Mach's principle, which for him had a tremendous value. This is understandable, for behind those ideas loomed the big open question which physics had not really answered: what is the source of inertia? And to which, let us say it straight away, general relativity would only provide an extremely partial answer. Mach's principle was not really included in Einstein's theory of gravitation and this would be a great disappointment for Einstein.

But even if all these ideas are very beautiful and the questions truly exciting, they are certainly not accepted by everyone. Many physicists are not touched by them; they are satisfied with a very precise description of these phenomena and are not troubled by the reasons behind them. It is well known that the goal of science is to find answers to "how" and not to "why." But is that real progress?

A THEORY OF THE GRAVITATIONAL FIELD

General relativity is a theory of the gravitational field. It is partly modeled after Maxwell's theory, which provides an answer to the following question: what is the electromagnetic field corresponding to a given distribution of charges or currents?

Let a source, a distribution of charges in Minkowski's space-time, be given. Maxwell's equations will allow us to calculate the resulting electromagnetic field. *All we have to do* is solve the equations, which is not all that difficult, for Maxwell's are linear differential equations. But these equations will not tell us what the motion of, say, an electron is in that electromagnetic field. If we want to know the path of a charged particle in a given electromagnetic field, we need a law of motion. More precisely, we have to define the electromagnetic force that the field applies to the charged particle, known as Lorentz's force f_l. We then apply the fundamental law of dynamics ($f = m_i a = f_l$) and, by integration, obtain the equations of motion of the charge (of inertial mass m_i) in the given field, for we are still working with differential equations. In this example we have ignored the electromagnetic field created by the electron, simply because we have implicitly assumed that it was extremely weak and hence negligible compared to the given field. But we could have very well taken it into consideration. What we have taken into account is the time that the electromagnetic field takes to travel in space-time; the Lorentz-Maxwell equations automatically take into account the speed of propagation of the electromagnetic waves.

Just as in Maxwell's theory, we may assume that a gravitational field, created by a distribution of (gravitational) masses that are the source of the field, travels in space-time. It is also necessary to suppose that a force, defined by the gravitational field and whose action is delayed (to allow for the time it takes to travel in space-time, with a speed that is generally that of light) acts at a distance on the particles. Finally, thanks to the fundamental law of dynamics, $f = ma$, the motion of the particle in the field can be obtained.

That is not too different from what occurs in the Newtonian theory of gravitation. In the same way that a source made up of charged particles creates an electromagnetic field, a particle of mass M_g, or more generally a distribution of (gravitational) masses, is the source of a gravitational field. This field acts on a particle of (gravitational) mass m_g by creating Newton's gravitational force ($f_g = m_g M_g G/r^2$). But, as observed earlier, Newton's theory acts in Newton's absolute space, not in Minkowski's, and the speed of propagation of gravitation is therefore infinite. The fundamental law of dynamics ($f = m_i$; $a = f_g$) will establish the connection between the force created by the gravitational field and the motion of the particle of (inertial) mass m_i subject to the field. The resulting differential equations enable us to calculate the motion of the particle in the gravitational field.

In a *relativistic* theory of gravitation, the gravitational field equations must be formulated in Minkowski's space. Then, just as a charged particle creates an electromagnetic field, a particle of a given mass, or more generally a distribution of (gravitational) masses, will be the source of a gravitational field. But in order to calculate the motion of a given test particle, it will also be necessary to know the action of the field on that mass, that is, the gravitational force exerted by the field on the particle. We will then obtain a sort of generalization of Newton's law of gravitation. Finally, the fundamental law of dynamics will establish the connection between the force created by the field and the motion of the particle. Notice that, in this context, a law of gravitation must take into account the gravitational interaction propagation time.

A theory along those lines is perfectly plausible, and it is a solution that Einstein had in fact attempted. He was not alone in trying this strategy: numerous (relativistic) theories of gravitation would be developed during the century. The principle of equivalence was not a priori assumed; the need for it came later, when the equations of motion had to be written.

In fact—and not surprisingly—the theoretical structure sought by Einstein bears some resemblance to that of Maxwell's theory of electromagnetism, in which the charges present create a field that acts on the charges in motion. In a rather analogous manner, for such is the model of field theory envisaged by Einstein, the masses present in the universe create a gravitational field to which the test particles are subject. These are called *test* particles because we assume that their masses are as faint as possible, so that they will not interfere with the given field. But the fundamental difference with Maxwell's theory was the framework in which Einstein worked to build his theory. It was through the curvature of space-time that the gravitational field would manifest itself. In 1912, Einstein could only catch a glimpse of precisely how this would be done.

Thanks to the principle of equivalence Einstein was able to treat gravitation as an acceleration field; he moved down one notch, going from a question in dynamics to one in kinematics, a question that special relativity allowed him to handle in a heuristic manner, just to see how it worked, until general relativity came along.

The principle of equivalence also allowed him to take one more step, an essential step that led Einstein to bend space, to free himself from Euclidian (or Minkowski's) space and adopt that of Riemann. As I emphasized above, the principle of equivalence rests on the fact that all bodies fall in the same way: the trajectory of test particles or corpuscles is independent of their nature; it is carved in the underlying space. We can then assume that gravitation is a property of space itself, or rather we can expect to build a theory in which gravitation would be a property of space.

Another possibility (other than a Minkowskian theory of gravitation) consists in proposing that space—that is, space-time—is the support of gravitation. But how can space represent gravity? The gravitational field varies from one point to another, so space must be different at different points; something in space-time has to change from one place to another. We need a space that is not everywhere the same. The physical spaces in which we generally work are the same throughout, here and there. They are like planes on which one travels and nothing changes; the laws defined on them do not change.

But what can be made to change from one point of space to another? Curvature is the simplest solution to this question. If we consider the set of possible surfaces, the plane represents one on which, from one point to another, nothing changes; any given place is equivalent to any other place. Physics will be the same here or there. But on a surface where curvature varies—a crumpled piece of cloth, a battered sheet of metal—then physics will not be everywhere the same. This difference can represent the gravitational field.

Imagine a plain on which we set a marble without rolling it. The marble will of course remain at rest. But if we repeat the experiment in a mountainous landscape, the marble will roll down in the direction of maximum slope, the direction along which the curvature of the ground is the greatest. Of course, this curvature is that of a surface and does not have the same properties as the curvature of space-time. But, even if imperfect, it is an interesting image. As I said before, there is no *right* image, no perfect representation of these concepts, only some less imperfect than others. When dealing with curvature, let us keep this one in mind and notice that the trajectories of our marble are inscribed in the ground, carved in space-time.

We thus see how a curved space-time can *represent*, how it can support, gravitation. The gravitational field will be a property of space-time; matter—the masses at the origin of the field—will therefore bend space-time, those masses that we used to call *gravitational* but which will now be the same as the *inertial* masses, thanks to the principle of equivalence. We still have to understand *how* curvature will depend on the masses that create the field and which field equations will determine the curvature of space-time, equations that will tell us in what kind of space we live (see chapter 5).

While the theory's field equations will determine the space-time that is a solution of the problem, we still have to invent the space-times among which that solution may be found. We need a supply of possible, satisfactory spaces. In fact, these curved spaces were invented by Bernhard Riemann in the middle of the nineteenth century. But the question is much older than that.

We surely remember the "parallel axiom" that we have all recited in a drone in elementary school and according to which "two parallel lines never meet." This is in fact a consequence of the fifth postulate in Euclid's famous treatise *Elements*, which is the basis of Euclidean geometry. In that fifth postulate, Euclid stated that through a point not on a line passes only one parallel, which is equivalent to affirming that "two parallel lines never meet." Euclid, and later many other geometricians, tried to develop plane geometry without resorting to this postulate, without using such a hypothesis. But was it really a hypothesis? Wasn't it possible, not just to do without it, but even to prove it? Countless proofs of the parallel axiom mark the history of Euclidean geometry, but they are all false, for it is really a postulate, that is, a hypothesis that must be stated but cannot be demonstrated. Then there was the question of the validity of the parallel axiom. Was it always valid, everywhere, and in every context? Is our geometry everywhere true, universal, and necessary? At the beginning of the eighteenth century, Carl Friedrich Gauss was convinced that it was not possible to prove the "necessity of our geometry,"[19] but he was alone to think so. One century later, Nikolay Ivanovich Lobachevsky created another geometry that he called "imaginary," in which the sum of the angles of a triangle was not 180°. This implies that in his peculiar geometry the parallel axiom was not true. Using astronomical data he then calculated the sum of the angles of a triangle formed by three stars and showed that it is in fact 180°, allowing for the errors in the observations. He thus pretended to have *proved* the parallel axiom.[20]

To prove, that is the problem. In fact, Lobachevsky did not prove the parallel axiom; at best, he verified it, for the measurements he used belong to astronomy, a natural science, and not to mathematics. He did show, though, together with Gauss, that the geometry question was not only a mathematical one. A distinction had to be made, as would be clearly done by Einstein, between *practical* and *axiomatic* geometry.[21]

Lobachevsky thus invented an imaginary geometry, essentially a geometry of the sphere, in which not only two parallel lines meet but the sum of the angles of a triangle is not 180°. His work is at the origin of the development of an important branch of mathematics: non-Euclidean geometry. The familiar theorems that we learned at school, such as Pythagoras's, are no longer valid in such geometries. In the middle of the nineteenth century, Riemann generalized those results and developed the basic techniques for building—and working in—non-Euclidean spaces possessing a curvature, the so-called Riemannian spaces. At the end of that century, Gregorio Ricci-Curbastro created an absolute differential calculus, tensor calculus, which we have already met and which is a mathematical method for working in Riemann spaces. Ricci and Tullio Levi-Civita then developed the methods that Einstein later used to work in those curved spaces and give form to his general relativity (see chapter 5).

Once the space-time that is the solution of a given problem is known, it is necessary to specify the motion of test bodies in a curved space-time. What is the equation of their trajectory? There is an immediate answer to this question: long before his theory of gravitation was correctly structured, Einstein knew that particles navigating in a curved space-time have no choice but to follow the geodesics.

In classical mechanics, the law of inertia stipulates that in the absence of any external force, a body or particle travels in a straight line with constant velocity, just like a billiard ball. If the particle is subject to an external force—gravitation, for instance, but it could also be an electromagnetic force it the particle happens to be charged—then the trajectory becomes complex. In the case of the Earth in the Sun's field, ignoring the other planets, the trajectory is an ellipse with the Sun at one of its foci.

Can we fix things up so that the law of inertia holds in all these cases? We sure can: we remove the gravitational forces and inscribe them, thanks to curvature, in space-time, and we set free the particles, which will follow the paths of the curvature—the geodesics. Thus, the law of inertia is generalized, the notion of force eliminated (integrated into the curvature of space-time), and the particles liberated. Another beautiful idea that requires further discussion.

The geodesic of a curved space is first of all the natural generalization of a straight line in a *flat* space. In Newtonian theory, *free* particles follow straight lines that they travel with constant velocity, like the balls in billiards (this is the law of inertia), while relativistic particles follow the geodesics of space-time. The great circles in the sphere generalize the straight lines of the plane; they are the geodesics of a spherical space. On an arbitrary curved surface, the shortest lines between two points are precisely the geodesics of that surface. In general relativity they are not the shortest paths but, paradoxically, the *longest* ones in proper time.[22]

Thus, rather than being subject to forces that compel them to obey gravitation, as in Newtonian theory, relativistic particles will simply follow the extreme paths of space-time. The particles in Newton's absolute space must obey his law of gravitation, while those in Einstein's curved space are free, free to follow the geodesics of space-time. As it will turn out later, this principle is a consequence of the field equations, and so it is already included in the "basic theoretical package" and need not be explicitly imposed.

We can already go a little farther and understand how it is possible to describe the man in free fall inside the elevator. The question we have to ask is, What is the structure of the space in which the man is falling, the space in which he does not feel, as we saw earlier, any gravitational field? It is of course a flat space. Physically, this reference system in free fall is known by the slightly barbaric name of "inertial *local* reference frame," a reference frame in which *gravitation is absent*. The theory's mathematicians prefer to use a no less barbaric name for this particular reference system: "the tangent space," tangent, that is, to the Riemannian manifold—actually, the osculating space. Because it does not represent any gravitation, this space is, unsurprisingly, that of special relativity—that is, Minkowski's space-time.

The existence of a gravitational field would manifest itself by the presence of curvature, of an acceleration field, and, conversely, the latter would indicate a gravitational field. But, as we have just seen, one could always *locally* (at one point of space-time) neutralize gravitation by letting oneself fall freely. In fact, gravitation, which is indistinguishable from acceleration, is relative to each observer. This is consistent with Einstein's wish to make acceleration a relative concept—it is not acceptable that the principle of relativity should hold for velocity (there is no absolute velocity; bodies only have velocity relative to other bodies) and not for acceleration. All the more so since the absolute character of acceleration is the basis of the inertial frames and hence of absolute space. And if Einstein wanted to get rid of Newton's absolute space—and he had no choice, if he were to uphold Mach's principle—he would not be able to keep the privileged role of inertial frames, for the collection of all inertial frames constitutes in a certain sense absolute space. Einstein thus developed a critical examination of the absolute nature of acceleration (which he wanted *relative*, like velocity). He also generalized the class of up-to-then admissible frames, that is, the inertial ones. It should be possible to write the equations of general relativity in any reference frame, whether accelerated or not. Hence, the description of physical phenomena was no longer restricted to the class of inertial frames (or of inertial observers, which is simply another way of putting it)—it was now *generalized*.

This legitimate refusal to restrict the admissible frames to the inertial ones led Einstein to undertake a critical evaluation of the physical meaning of coordinates. He found himself caught in a sort of chain of questioning. His criticisms of absolute space led him to that of inertial systems, and then to reflecting on the meaning of coordinate systems. In short, he aimed at an all-out generalization that left few independent concepts in physics standing. General relativity is a very economical but not very self-revealing theory.

The question was then to generalize the principle of relativity, that is, to find the equations which, taking into account gravitation and without excluding acceleration, can be expressed in arbitrary, though not necessarily inertial, coordinate systems which may be in acceleration with respect to each other. That is one of the main axes of the general theory of relativity.

THE PRINCIPLE OF COVARIANCE

Yet another principle? It is, rather, a tool, a mathematical tool that expresses the invariance of the laws of nature. It is simply a matter of being able to express equations (provided we know them) in *any* coordinate system and to know how to use them. Let us recall the statement of the principle of relativity in the special theory: the laws of physics that are valid in a given inertial frame must also be valid in any other inertial frame; or, put another way, these laws must be invariant with respect to the Lorentz transformations (those transformation laws that allow us to go from one inertial frame to another); or still, in more *professional* language, the laws of physics must be Lorentz *invariant*.

The principle of covariance expresses the essential idea according to which the laws are intrinsic and must therefore have the same inherent form, even if their expression will vary depending on the coordinate system in which we choose to work. A particular mathematical tool, the theory of tensors (box 4), is a formalization of the way the components of the objects involved are transformed under a change of reference frame or coordinate system. The principle simply expresses the fact that these objects are intrinsic, that their components must be invariant. Finally, we must specify the reference frames for which this invariance must hold and the transformations under which these objects will remain invariant. In the context of the classical theories the latter are the Galilean transformations, and in special relativity, the Lorentz transformations. As for general relativity, these were all conceivable transformations. Einstein then postulated the general covariance of equations, a principle of general relativity.

Likewise, all coordinate systems, all reference frames must be susceptible of being used, for it must always be possible to absorb (or to generate) a gravitational field through an acceleration

Box 4. Tensor Calculus

Tensor calculus is the tool of covariance. Thanks to a system of indexes, we can identify, for geometrical objects such as vectors, their components with respect to a given axis of a particular coordinate system. A vector A will then have a component A^1 with respect to the axis x_1, A^2 relative to x_2, A^3 relative to x_3, and A^4 with respect to the axis x_4 (which represents time). In short, there will be four components, of value A^μ relative to the axis x_μ (where μ takes on the values 1, 2, 3, 4). We can employ at the same time a second coordinate system x'_ν (x'_1, x'_2, x'_3, x'_4) and we shall call A'^ν the components of the same vector A (with ν also taking on the values 1, 2, 3, 4). A vector is a geometrical object (and therefore invariant as such) whose components are the projections on particular axes (x_μ or x'_ν, whichever the case). Then, the tensorial machinery will allow us to go automatically from one representation to the other, accounting for the changes in the coordinate system and in the components of the vector. All we need to know is the transition "matrix," denoted M^μ_ν, which performs the automatic conversion from one coordinate system to the other. There are other, more complex mathematical objects called tensors that have more components ($4 \times 4 = 16$ for a tensor of rank 2) and which are transformed like vectors, in a way slightly more involved but just as automatic.

field. We must be able to write the equations of gravitation, that is, those of general relativity, with respect to any accelerated reference frame: the covariance must be general. That is the *general* character of *relativity*.

However, we may look at things differently: from an "invariant" perspective. I have already emphasized that it was a misnomer to call Einstein's theories "theories of relativity," for they were in fact absolute theories whose *relative* character only appeared when the physical magnitude under consideration was projected on a particular reference frame; hence "relative" to that reference system. In other words, we are dealing with intrinsic objects that we wish to project on some coordinate system in which it will be easier to work. But it is not impossible to project them on another reference frame or system of coordinates with a different acceleration field—the theory of tensors will take care of this *automatically*. In fact, the necessary calculations for the projection on a given coordinate system or for the transformation of the components of the intrinsic object in question from one frame to another is all guaranteed by the tensor calculus. In fact, these questions are similar to those in industrial design or architecture. In order to obtain the exact dimensions of the object we want to manufacture or the building we wish to construct, we project the structure—the shape—on the plane and do this from each of its sides. If we change the representation, if we wish to draw the building in perspective, for instance, it is natural to expect that some mathematical tool will permit us to carry out this transformation, this change in the point of view. But such a tool will be just that, a representational tool, while the building itself remains unchanged, identical to itself. In this simple example we can see more clearly how problematic the relative character of general or special relativity is. It is only a relativeness of the description; the essence of the theory is the invariance, as I have already emphasized in the previous chapter.

NOTES

1. Einstein, 1950, p. 13.

2. Einstein, 1919a, manuscript 2-070-23. It is precisely in 1907, while he was writing a program for a relativistic theory of gravitation in his review of special relativity, that this idea occurred to him. This is what he says in a 1919 text that has remained until today in manuscript form, the Morgan manuscript.

3. Einstein, 1919a, manuscript 2-070-23-24.

4. Ibid.

5. R. von Eötvös, 1874.

6. Cavendish had obtained a first torsion balance from his friend John Michell, whom we shall meet in chapter 8. It appears that the torsion balance was invented independently by Michell and Charles

Augustin Coulomb, who had built a balance based on the same principle and which allowed him to establish that the electric force is proportional to $1/r^2$ ("Coulomb's law"). To learn more about Cavendish, see Jungnickel and McCormmach, 1999. Cf. Gillmor's book, where the priority question is discussed: Gillmor, 1971, pp. 163–165.

7. Hoyle et al., 2001.

8. In this regard, cf. Williams et al., 2001.

9. Einstein, 1907, *CPAE*, vol. 2, p. 476.

10. Mach, 1883.

11. Cf. Mach, 1904, pp. 216–235. Cf. also my analysis of these questions in the *Cosmologie* chapter, Eisenstaedt, 1993.

12. Ibid., p. 222.

13. Ibid., p. 230.

14. Einstein, 1916. German original in *CPAE*, vol. 2, p. 286; English translation from Lorentz et al., 1923, p. 113.

15. Foucault, 1851.

16. Einstein, 1912, *CPAE*, vol. 4, p. 177; English translation vol. 4, p. 128.

17. Einstein to E. Mach, 25 June 1913, *CPAE*, vol. 5, p. 531.

18. Einstein to E. Mach, 2 December 1913, *CPAE*, vol. 5, p. 584.

19. C. F. Gauss to H. W. Olbers, 1817, quoted in Boi, 1995, p. 75.

20. In this regard, see Rosenfeld, 1988, pp. 206–208.

21. A question also raised by Poincaré and many other physicists, such as Karl Schwarzschild (Schwarzschild, 1900). A fundamental question to which we shall return in chapter 6.

22. This is due to the fact that the terms of the metric are not all positive, which implies that two distinct events may be at a (proper) distance zero from each other. . . . In this regard, cf. Hakim, 1994, p. 174.

CHAPTER FIVE

The Birth of General Relativity

IN 1907 EINSTEIN was only twenty-eight years old. He was still a young man, a humble technical officer at the Swiss Patent Office in Bern, a position he only accepted for want of anything better. He had not really been a brilliant student, perhaps because he only applied himself to subjects that interested him. His former teachers did not have a very high opinion of his potential and were insensitive to his vocation and his passion; he was a passionate, revolutionary young man. Which *Herr Doktor Professor*—Albert had always studied in countries of German culture—would hire and support a young man already labeled as a lazybones and who had the reputation of being a rebel? A few years earlier his worried father had even taken the liberty of writing, very respectfully, to Wilhelm Ostwald, a professor at the University of Leipzig, asking him to support Albert's candidacy for a position at the university—to no avail.

Einstein did not really despise those boring and rather sad old Herr Professors. Instead, he seemed to suffer in silence for the contempt and the neglect he had to endure, but he would soon take his revenge: Albert, the rebel, the revolutionary, knew already that he had put his finger on the solution of some of the most important and profound problems in physics. Physics was everything to him, to the point of sacrificing to it—both he and Mileva, his wife—their daughter Lieserl, whom shortly after her birth they abandoned to a foster family, as it is discreetly called nowadays, from Mileva's Serbian village. A sacrifice to *his* vocation, to *his* career: a sacrifice that must have been an open wound in the life of this young woman with a passion for physics as intense as her love for this physicist that she, well before anyone else, had recognized as a genius. The couple would never recover from that blow. It was a sacrifice that she granted

Figure 5.1. Albert Einstein. (© AIP Emilio Segrè Visual Archives)

him not without regret and which may explain why, when they got divorced at the beginning of the 1920s, Albert gave Mileva the Nobel Prize money he had just received.

Albert already had three articles under his belt, the three most important articles of the century, and yet his career, to which he had sacrificed everything, had still not taken off. But things were beginning to change, and Einstein was not surprised to receive in the fall of 1907 a letter from Johannes Stark, the editor of a prestigious German journal, the *Jahrbuch der Radioaktivität und Elektronik*, asking him to write a review article on the theory of relativity, which was already a kind of recognition. A few days later another letter arrived, this one from Hermann Minkowski, which must have given Einstein particular pleasure. Let us not forget that in the years 1897–1900, while he was a student at the E.T.H. (the Swiss Federal Polytechnic), he had Minkowski as his mathematics professor. Of these classes, the student did not have any fond memories and neither did the teacher, who considered his pupil a "lazy dog,"[1] a not very diligent student who did not work enough and who failed to master the mathematical techniques he was taught.

Minkowski had since left Zurich for the University of Göttingen, where he was David Hilbert's colleague. He was in charge of a seminar in which, he wrote to Einstein, "we also wish to discuss your interesting papers on electrodynamics" and, he added, "I was in Zurich recently and was pleased to hear from different quarters about the great interest being shown in your scientific successes."[2]

We have already seen in previous chapters how deep Minkowski's interest in special relativity was. It led him to reformulate the theory in a revolutionary four-dimensional formalism that Einstein would be very reluctant to accept.[3] But he eventually appreciated the significance of Minkowski's approach and incorporated it at the very heart of his program as an essential element.

Let us return to the review article he was currently writing, which presents relativity in a more synthetic and pedagogical fashion than did the 1905 paper.[4] In this new article, Einstein goes quickly over the reasons for special relativity, recalling the impossibility of detecting the motion of the earth with respect to the ether, attacking Lorentz's hypothesis of the contraction of lengths, referring for the first time to the Michelson-Morley experiment that rendered the ad hoc Lorentz's postulate superfluous, and calling the latter "an artificial way of saving the theory."

He mentions, of course, the collection of results of the past two years of work, in particular those in his 1905 article on "energy's inertia," that is, "$E = mc^2$."[5] He insists on the fact that "during the radioactive decay of a substance, vast quantities of energy are being released,"[6] and he quotes Planck, according to whom radium would liberate an amount of energy corresponding to a decrease in mass of 1.41×10^{-6} mg per hour.

But if mass is energy, then energy is also mass, and in particular inertial mass. And if, according to the equivalence principle, inertia is *equivalent* to gravitation, then inertial mass is nothing but gravitational mass. Einstein is thus led to "assume that radiation enclosed in a cavity possesses not only inertia but also weight."[7] "$E = mc^2$" is therefore one more step, one more bridge starting at special relativity and reaching toward gravitation. This is the program Einstein will develop in the fifth and last part of his 1907 article.

The title of this last part is in fact "the principle of relativity and gravitation" and it unfolds in several stages.[8] First of all, Einstein tackles the problem of the generalization of the principle of special relativity to accelerated systems, a question that is "to some extent accessible to theoretical treatment." This provides him with a foothold: thanks to the principle of equivalence, which states that there is no difference between "a uniform gravitational field" and "a uniformly accelerated system," it is possible to go from one to the other. Thus, in the 1907 paper, Einstein drew up the plan of the battle he would fight for the next eight years and which would allow him to build a theory of gravitation compatible with special relativity.

A CAVALIER ARGUMENT

After formulating his research program, Einstein takes the first step: the behavior of clocks in an accelerated system and, consequently, in a gravitational field. But for the time being he only has at his disposal 1905 relativity restricted to Galilean systems, which are, by definition, nonaccelerated. The problem is therefore how to deal with accelerated systems without really having the right to do so. Einstein employs two reference frames, both equipped with identical clocks: S, a frame assumed to be at rest (nonaccelerated) and an accelerated frame Σ. To go from S to Σ, a step that special relativity does not allow, he must find a subterfuge. He then imagines a third system, S′, which is supposed to coincide with Σ, but it is nonaccelerated. Thanks to (special) relativity (there is yet no general relativity), he is then able to go from S to S′ and then, by *sliding* from S′ to Σ, go from S to Σ and have a way of predicting the behavior of an accelerated clock.

It is not a very complex calculation but rather a subtle and quite annoying argument requiring the manipulation of three different reference frames at the same time. One has to deal with clocks, local times, signal exchanges, and the definition of simultaneity. The method used here by Einstein is typical of those proofs requiring going from one train to another while keeping an eye on the stationmaster on the platform but without forgetting the age of the driver. We can easily lose our bearings! And in the end, we come out with an uneasy feeling: isn't there something wrong? In fact, it is true that something is not right, but we are moving forward, and that is what we should expect from a *heuristic* argument. Making progress is what matters, for Einstein's laborious calculations lead to the formula spelling out the influence a gravitational field has on a clock; it is known as the *Einstein effect*: gravitation slows down clocks. Here is how Einstein puts it: "We may say that the process occurring in the clock—and more generally, any physical process—proceeds faster the greater the gravitational potential at the location of the process taking place."[9]

In fact, his wording is problematic and today no one would state it that way, for the proper ticking of a clock, or the vibration of an atom (which expresses the idea almost perfectly—cesium clocks, for instance) is an *invariant*: the proper frequency of an atom is independent of the gravitational field in which it is embedded.

This seems to be a contradiction, a paradox. How can the clock be really slowed down without its proper frequency being changed and without the possibility of locally measuring the variation? It is a question of definition: the proper frequency is the proper frequency, and that's all. When I look at my watch, I measure the time in my life, and everybody else does the same. But I can compare my proper time with that of my twin, or compare the frequency of the atom I'm carrying with that of an atom on the Sun or a star. The comparison of times or frequencies will tell me something about the variation in the structure of space-time, but we shall remain twins and the two atoms will remain identical.

Einstein's formulation is not completely right, and it is even incorrect. He is caught between two interpretations, and this will later force him to revise his formulation and arrive at one that is conceptually correct. Discovery often takes some side streets that are not necessarily the shortest or the cleanest!

Einstein goes on and speaks of "the existence of 'clocks' that are present at locations of different gravitational potentials and whose rates can be controlled with great precision; these are the producers of spectral lines. It can be concluded from the aforesaid that the wavelength of light coming from the sun's surface, . . . is larger by about one part in two millionth than that of light produced by the same substance on earth."[10]

This simple calculation—where Einstein, contrary to a persistent myth, is already concerned with experimental verification—marks the birth of the Einstein effect or, more precisely, the shift of atomic lines in a gravitational field. This effect has now been accurately verified (see chapter 9).

He then tackles another fundamental question, "the influence of gravitation on electromagnetic processes," and after writing Maxwell's equations in an accelerated frame, he manages to show that the speed of light is a function of gravitation. Again that same offhand argument: Einstein works with something that does not belong to him, at least not yet. But there is no gain without risk. Beside, his conclusion may (and must) appear bizarre: is he not working in the context of special relativity, where the speed of light is, by one of his principles, constant? His calculations are somewhat shaky and should not be taken too seriously; they indicate a lead, a possible or probable effect that needs rationalizing. But, even if Einstein is fishing for a solution, he knows that his investments are sound. He is simply using heuristic reasoning and taking, when necessary, some liberties with his principles.

Leaving aside for a moment the principles of special relativity, let us assume the speed of light to be a function of the intensity of gravity. As if it traveled through a crystal or other refringent medium, the speed of the light wave will vary depending on the strength of the gravitational field, and the wave will therefore be deflected according to Huygens' law. This optical analogy, where the light wave appears subject to a refringent medium represented by the gravitational field, would be used by Einstein in a later article on the deflection of light in a gravitational field, and it would be employed again in the 1920s in order to better understand, to better explain, and especially to better "see" this kind of phenomenon.

This somewhat fussy discussion left at the time, as it leaves even today, an ambiguous impression: to what extent must we believe those pointless calculations? Yet, in one sweep, and not happy with clarifying the main ideas of his still embryonic theory, Einstein has just pointed to two or three tests for his theory. The line shift and the deflection of light rays would be crucial tests for general relativity. Despite the difficulty in setting them up, these tests, together with that of the changes in the perihelion of Mercury would soon constitute, and for a long time, the only support of the theory. We shall come back to them in future chapters (see chapters 7, 8, and 9).

A JEWISH PHYSICS?

On Christmas Eve 1907 Einstein wrote to Conrad Habicht, an old friend from Bern and one of the three members of the Olympia Academy, a letter with which he enclosed the proofs of his article: "In the months of October and November I was very busy with a paper—partly a review and partly a presentation of new material—concerning the Relativity Principle. At the moment

I am working on a relativistic analysis of the law of gravitation by means of which I hope to explain the still unexplained secular changes in the perihelion of Mercury" and, he added as a postscript: "But for the moment it doesn't seem to work."[11]

Apparently Einstein succeeded only in asking—as usual—the right question: the one concerning the advance in the perihelion of Mercury, a phenomenon that Newton's theory could not explain; a real anomaly that would not be explained by general relativity until 1915 in what would become the *first classical test* of his theory.

Hence, in a few months, Einstein had laid down not only a good part of the foundations of his theory but also the physical logic of the three tests that would support it for fifty years.

Encouraged by Alfred Kleiner, a professor at the University of Zurich who had supervised his dissertation on a new definition of molecular dimensions, Einstein applied for the post of *Privatdozent*, an appointment without pay but nonetheless a stepping stone for a position at the University of Bern. But the head of the department of physics blocked his appointment, arguing that Einstein's article on relativity was "incomprehensible."[12] In the spring of 1908, Kleiner advised him to try again, and this time his candidacy was accepted. Einstein had to teach a seven o'clock morning class that attracted only four students, his friend Michele Besso among them. Soon after, Kleiner pressed him to apply for a newly opened position at the University of Zurich. But there was another candidate, one who was well known and whose appointment was virtually guaranteed: Friedrich Fritz Adler, son of the founder of the Austrian Social Democratic Party and who had some close friends among the members of the hiring committee. Adler, well aware of Einstein's credentials and not really interested in the position (he wished to pursue a career in politics), wrote to the hiring committee in support of Einstein's candidacy. But the latter had another handicap: his Jewishness. The final report could not be clearer:

> These expressions of our colleague Kleiner, based on several years of personal contact, were all the more valuable for the committee as well as for the faculty as a whole since Herr Dr Einstein is an Israelite and since precisely to

> the Israelites among scholars are ascribed (in numerous cases not entirely without cause) all kinds of unpleasant peculiarities of character, such as intrusiveness, impudence, and a shopkeeper's mentality in the perception of their academic position. It should be said, however, that also among the Israelites there exist men who do not exhibit a trace of disagreeable qualities and that it is not proper, therefore, to disqualify a man only because he happens to be a Jew. Indeed, one occasionally finds people also among non-Jewish scholars who in regard to a commercial perception and utilization of their academic profession develop qualities which are usually considered as specifically 'Jewish.' Therefore neither the committee nor the faculty as a whole considered it compatible with its dignity to adopt anti-Semitism as a matter of policy and the information which Herr Kollege Kleiner was able to provide about the character of Herr Dr Einstein has completely reassured us.[13]

It was a judgment that, even if written with the purpose of helping Einstein's cause by clearing him of all suspicion, remains nevertheless a typical example of a virulent anti-Semitism.

Einstein's Jewish background is a rather surprising issue that surfaces a number of times and is related to his work habits. In a letter to Lorentz at the end of 1907, Sommerfeld wrote: "We are now anxiously waiting for your views on Einstein's papers. Despite their being brilliant, I still find something almost unhealthy in that incomprehensible and nonintuitive manipulation of dogmas. An Englishman could hardly have proposed such a theory; it is perhaps, as in Cohn's case, the expression of the abstractly conceptual style of the Semite. Let us hope that you will succeed in filling that brilliant skeleton with some true physics."[14]

Sommerfeld had not yet read the 1907 article, which Einstein would send him as soon as it appeared at the beginning of 1908, and therefore his remarks were aimed at special relativity: perhaps not so much at the theory itself as at its presentation, because it contained no references to articles or experimental results and was supported only by its two principles. Sommerfeld's views were no doubt prompted by Einstein's fierce discussion of speeds higher than that of light, which were a priori ruled out by special relativity.

Much later, in the course of a lecture, Max Born would define "Jewish physics" as a particular effort "to get hold of the laws of nature by thinking alone." To which Einstein would reply that "Jewish physics was not about to disappear." Born's own reply was to the effect that he had "always well understood and much appreciated [his] good Jewish physics."[15]

Thus, it was despite his Jewishness that in January 1909 Einstein was appointed to the University of Zurich. Two years later he accepted a position at the German University of Prague. It is worth noticing that despite the slightly bohemian side to his character, Einstein would make all the right career moves. He did not hesitate to play the University of Utrecht, which had offered him a post, against the E.T.H. in Zurich, where he had been a student and which would appoint him for ten years in 1912, after his stay in Prague.

Five years after special relativity, Einstein's stock was rising. Planck, a member of the commission that supported Einstein's appointment at Prague, wrote in its 1910 report: "In its breath and profundity the revolution in the physical world-view occasioned by [the principle of relativity] can only be compared with that brought about by the introduction of the Copernican world-system."[16] Today it may be hard to understand the need for such an obvious statement; however, the relativistic revolution was so radical that to hear it back then was far from obvious. It was also because Einstein had come a long way from an obscure beginning, and his trajectory appeared amazing, as the extent of Minkowski's change of opinion demonstrates. In a matter of two to three years, Einstein moved from one of the last places to the very first position in the world of theoretical physics—an achievement that we may find touching because it was for him (and hence, a little for any of us as well) a revenge against the *apparatchiks* and the heirs, and an accomplishment which probably explains in part the glory that would later be his. Einstein was not the typical heir of the university world; he did not receive any help during his studies, which he completed in solitude and besieged by difficulties. It was solely to his own merit that he owed his accomplishments and his success. But it was also due to Einstein's bold attitude in taking risks: his first priority was not his career but realizing what he believed in. He risked it all and his success resembles that of a great artist who narrowly missed not being recognized. What is more, the difficulties he was supposed to have had as a child—that he was almost intellectually retarded, according to some—made of him a sort of ugly duckling who, tardy in coming out of his shell, would metamorphose into the prince of the starred universe that his theories revealed. A fairy tale in the tough world of theoretical physics.

EINSTEIN AT WORK

Einstein did not publish anything else on gravitation until 1911. His post at the University of Zurich involved a heavy teaching load, which he undertook with great pleasure. Despite little teaching experience, he got along very well with his students and encouraged them to take an active part in the class. He taught mechanics, the kinetic theory of gases, the theory of heat and statistical mechanics, and even electromagnetism. His teaching duties did not prevent him from doing some research, in particular on the question of radiation and on opalescence. And he did, of course, participate in discussions on the interpretation of relativity; in particular in a very interesting one concerning a rigid rotating disk, a question that lay between special and general relativity.[17]

The origin of this controversy was a proposed definition of the concept of rigidity in special relativity due to Max Born and which led to the question of the possible deformation of a solid. Paul Ehrenfest, a physicist from Leyde and one of Einstein's best friends, got involved by putting his finger on an annoying paradox. It was known that special relativity entailed a phenomenon called *contraction of lengths*, according to which a ruler in motion (i.e., in a direction parallel to the ruler) will be shorter that an identical ruler that remains at rest. On the other hand, the width of the ruler, being perpendicular to the motion, is not contracted. From this follows that the radius of a rotating disk, which is always perpendicular to the motion, will undergo no contraction. But the elements perpendicular to the radius, and hence tangent to the circumference, will be affected by the contraction, and so will the circumference itself. The contradiction stems from the calculation of the length of the circumference, which equals $2\pi r$ (in Euclidean geometry) but which should be a little smaller due to the contraction. Where is the error? I will not enter here into the details of the debate, whose eventual resolution involved abandoning the notion of rigidity in the theory of relativity and eliminating all reference to distance in favor of proper time. In particular, Von Laue showed that the concept of rigidity violated the fact that no signal can have a speed exceeding that of light. He pointed out that an absolutely rigid body would be able to transmit a signal with infinite speed: if we (rapidly) move one end of a steel ruler, the signal of this motion appears to travel instantaneously from one end of the ruler to the other, which is now unacceptable from the standpoint of special relativity, for all information takes a certain time to be transmitted.

In the spring of 1911 Einstein returned to the question of the effect of weight on the propagation of light, in part because his 1907 presentation of it did "not satisfy me"[18] but also and especially because he realized that such an effect was "capable of being tested experimentally."[19] He had just found a way to display the deflection of light produced by a gravitational field, an effect that he had hinted at back in 1907.

In his 1911 paper, Einstein calculated the deflection of light beams using Huygens' principle (which is commonly used to calculate the refraction of light on a crystal): a wave traveling in a gravitational field undergoes a deflection. Just as in 1907, the principle of equivalence allowed him to replace the gravitational field with a uniformly accelerated reference frame and so to determine the effect of gravitation on the local physics: the deflection of the light rays and an atom's line shift. Despite the correctness of the result, his calculations lacked rigor, for he was forced to assume that the speed of light varies from one place to another, thus violating one of the basic tenets of special relativity—the constancy of the speed of light. Such an ambiguous method would trouble the minds and the calculations of those who, less flexible and less capable than Einstein in finding the right path, would not use it so skillfully.

But we must not confuse heuristic considerations with normal and established practice. During a theory's initial stages, the right methods are being sought; they evolve and are refined, if I may put it that way, only gradually, long after the birth of the theory in question. And it is not surprising that those ancient methods continue to haunt textbooks and minds. Physical theories are living entities; they are not only born but also go through childhood, adolescence, old age, and death, for other reasons. The fact that those calculations were more or less correct, not quite perfect, was simply a detail. What really count are the arguments on which they are based, the qualitative aspect, the very existence of effects; the chain of calculations will be reconstructed later, with steadily improving precision and neatness. Beside, Einstein only tepidly defended his reasoning and his calculations; it is after all only heuristics and, as he put it in the conclusion of his paper, "the considerations here presented may appear insufficiently substantiated or even adventurous."[20] But, as so often happened, his intuition was the right one, and even if his calculations were ambiguous and cavalier, they provided a foretaste of the effect and the correct order of magnitude.

After having calculated the—very weak—extent of the deflection, he explained how to go about observing it. The basic idea is that the trajectory of the light rays emitted by a fixed star will be slightly deflected from a straight path if those rays pass near a large body. The latter could well be the Sun, but its own brightness conceals the stars close to the solar disk whose images would be most affected by the phenomenon—except during an eclipse, when the hidden sun permits a view of the stars in its vicinity. In that situation, the field of stars around the Sun could be photographed and compared to its image when the Sun is not present:

> A ray of light going past the sun would accordingly undergo deflection to the amount of $4 \times 10^{-6} = 0.83$ seconds of arc. The angular distance of the star from the center of the sun appears to be increased by this amount. As the fixed stars in the parts of the sky near the sun are visible during total eclipses of the sun, this consequence of the theory may be compared with experience. With the planet Jupiter the displacement to be expected reaches to about 1/100 of the amount given. It would be a most desirable thing if astronomers

> would take up the question here raised. For apart from any theory there is the question of whether it is possible with the equipment presently available to detect an influence of the gravitational fields on the propagation of light.[21]

Many astronomers would take up the challenge, first among them, Erwin Freundlich, but it would not be an easy task. We shall look in detail at those observations, which will culminate in 1919, in chapter 8.

A MURMUR OF INDIGNATION . . .

In the fall of 1911 Einstein was invited to the Solvay Conference on kinetic and molecular theories along with all the great figures of the small world of theoretical physics, notably Hendrik A. Lorentz, Max Planck, Jean Perrin, Paul Langevin, and Henri Poincaré, to name only those who were close to him in one way or another.

At around that time he wrote a review article that drew from different sources, in particular from the recently published work of von Laue on special relativity and also, and this is a significant step, from Hermann Minkowski's work which, as we have seen, Einstein struggled to digest.

He had gone back to work and tried, slowly edging forward, to draw the consequences of his 1907 hypotheses, and particularly of the principle of equivalence, by working in a constant acceleration field. In a series of papers he formulated a static and relativistic theory of the gravitational field. In the introduction to his first article, he returned to his analysis of the rotating disk and expressed his doubts that the law of geometry should hold in this case.

But it was Max Abraham who, borrowing Einstein's idea of a variable speed of light, would take the next step. At the beginning of 1912, he introduced Minkowski's line element of space-time in order to generalize to dimension four Poisson's equation of the gravitational field.[22] Einstein kept abreast of Abraham's work, which he was quick to condemn for not including the principle of equivalence. He was himself working on a theory of the gravitational field where the (variable) speed of light also represented the field. He then stated the important geodesics principle: the equations of motion of the particles in the field are simply the geodesics of the proposed space-time. But he still resisted adopting Minkowski's reformulation of the special theory, even if it was the route toward the generalization he was after. He was not far now from bringing about the great revolution of the curvature of space.

In March 1912, Einstein wrote to Besso: "Lately, I have been working like mad on the gravitation problem." And, after having explained his latest results, he returned to the subject of rotation: "You see that I am still far from being able to conceive of rotation as rest! Each step is devilishly difficult, and what I have derived so far is certainly still the simplest of all. Abraham's theory has been created out of thin air, that is, out of nothing but considerations of mathematical beauty, and it is completely untenable. How this intelligent man could let himself be carried away with such superficiality is beyond me. To be sure, at first (for 14 days!) I too was totally 'bluffed' by the beauty and simplicity of his formulas."[23]

The small world of theoretical physics was not ready to support his work on gravitation or even the path down which he pursued it. While Planck was definitely not seduced by the theory, von Laue, who had nevertheless very favorably greeted special relativity, rejected the principle of equivalence and accepted neither Abraham's nor Einstein's work. The latter complained to Besso: "The fraternity of physicists behaves rather passively with respect to my gravitation paper. Abraham seems to have the greatest understanding for it. To be sure, he fulminates against all relativity in *Scienza*, but he does it with understanding."[24]

In the summer of 1912, he left Prague after a short stay there and returned to Zurich, where he remained as a professor at the Polytechnic (E.T.H.) for two years before being appointed to a position in Berlin. In Zurich he was reunited with his friend Marcel Grossmann, now a professor of mathematics at the Polytechnic. Einstein sought Grossmann's help to understand Ricci's and Levi-Civita's absolute differential calculus and to learn all about tensors and curved spaces. Grossmann's assistance proved crucial in allowing Einstein to confront those dry mathematical concepts he had up to then always avoided. He said as much to Sommerfeld: "I am now working exclusively on the gravitation problem and believe that I can overcome all difficulties with the help of a mathematician friend of mine here. But one thing is certain: never before in my life have I troubled myself over anything so much, and I have gained enormous respect for mathematics, whose more subtle parts I considered until now, in my ignorance, as pure luxury! Compared with this problem the original theory of relativity is child's play."[25]

During those years Einstein worked frantically and more than once he lost his way after believing he had finally understood. Regarding his work on gravitation, he confessed to a former University of Zurich student: "I work *like a horse*, even if the cart does not always move very far from the spot."[26] And, a little later, apologizing for his long silence, he wrote to Ehrenfest that "my excuse is the frankly superhuman effort I have invested in the gravitation problem." But he added, enthusiastically: "I am now deeply convinced that I have gotten the thing right, and also, of course, that a murmur of indignation will spread through the ranks of our colleagues when the paper will appear."[27] Weary, he did not realize that his struggle was far from over and that he would still have to devote considerable time and effort to the gravitation problem. Worse still, he was not even on the right track for he had just given up, for the wrong reasons, his principle of covariance. His "Outline of a Generalized Theory of Relativity and of a Theory of Gravitation"—actually an initial version of general relativity—which had just appeared, did not create any furor among the sparse ranks of theoretical physicists. Incomprehension and indifference best describe its reception.

An essential element—in fact a principle—was still missing, and he would borrow it from Mach: Mach's principle, what a great idea! He was delighted. In his student days in Zurich and Bern, in those days of the Olympia Academy, he had discussed Mach's *Mechanics*[28] with Besso and Solo. The time had now come to include it in his plan. Mach's principle became a kind of obsession for Einstein. I have already explained the idea in the last chapter. As was often the case with Einstein, it was a simple, subtle idea concerning the physical reasons for inertia, which for Newton were inseparably linked to absolute space. Einstein refused to ascribe any physical meaning to Newton's absolute space, and so far he had been able to make progress without resorting to such a concept. Absolute space, nothing but a ghost!

A NETWORK OF PRINCIPLES

At this point, let me emphasize the total consistency of all the ideas and principles of the budding theory, a consistency whose corollary and limit was the immense freedom of thinking that Einstein had created for himself.

At the basis of this network, of this program, there were first of all, the initial principles, already classical, and foremost, of course, the principle of equivalence, the cornerstone of the whole building, because it enabled the unification of gravitation and inertia. Out with absolute space: matter itself was supposed to give the universe its particular structure, even in its most remote corners, including infinity. Mach's principle and the principles of covariance and general relativity pointed in that same direction. To those initial principles, which defined the framework of the theory, Einstein added a second layer of principles that were no less important. First, the all-powerful principle of geodesics with its far-reaching implications, for not only does it enjoin particles to follow the shortest pathways of space-time, but it also represents a kind of extension of the law of inertia that marks the relinquishment of the notion of force. Free test particles are no longer subject to a force but must simply closely *follow the relief*, the curvature of space-time created by gravitation. Next, the principle of conservation, which requires an equation of conservation of energy. And finally, the principle of correspondence, which demands that the new theory produce at least as many results as the one it seeks to replace. Thus, at the first degree of approximation, one should be able to recover all the results of

Newton's theory of gravitation: the new theory should not have fewer applications than the old one, nor, of course, less interesting results.

Even if their number is surprisingly high, these principles did not completely determine the theory and its field equations any more than did the mathematical structure that took Einstein years of relentless work to choose and build—a mathematical structure that would underlie the edifice and represent as much as possible his conceptual choices. His task was then to build a theory that could account for the Newtonian facts as well as be able to predict new ones.

A starting point would be to *generalize* the equations we already have—Newton's. The Newtonian equation that is the closest to what we are looking for is Poisson's, which permits us to calculate the so-called gravitational potential as a function of the mass distribution, that is, of the a priori given density of matter. Basically, Poisson's equation allows us to regard Newton's theory as a theory of the gravitational field. The right-hand side of the equation contains the source of the field, and the left-hand side, the mathematical operator applied to the gravitational potential U, which is the unknown function

$$\Delta U = 4\pi G\rho.$$

Poisson's equation, which should be read from right to left, enables us to calculate the gravitational field arising from the gravitational potential U by simple mathematical operations—differentiations.

We must therefore generalize Poisson's equation to render it compatible with special relativity, in particular by introducing c and making it time-dependent. This simple idea was explored by many theoreticians in one form or another, either directly from Newton's $1/r^2$ law, as Poincaré did starting in 1905, or from Poisson's equation, as Abraham and Einstein did in 1912. In essence, the idea is to build a theory of gravitation on a flat space—that is, on a Minkowski space—using the theory of electromagnetism as a model. For a number of reasons, ultimately related to observation, these *Minkowskian* theories of gravitation did not pan out, thus forcing Einstein, much to his satisfaction, to turn to a theory built on a curved space-time.

The structures of the field equations were similar to that of Poisson's equation: the sources act on the gravitational potential through an operator; a generalization, in a certain sense, of Poisson's mathematical *machinery*. This is the logic that Einstein would try to generalize to a curved, or Riemannian, space. On the right side of the equation were the sources of the field, the phenomenological description of the matter in space—the matter tensor; on the left, was an operator, a mathematical tool defined on a Riemann space that acted on the gravitational potentials. All of this defined a set of differential equations that specified the gravitational problem, the solution of which was essentially a particular Riemann space representing the gravitational field corresponding to the given source. Thus, to each source, to each distribution of matter, there corresponded a Riemann space that represented the associated gravitational field. There remained the question of finding the *right* operator. In the years 1912–1915, that would be one of the most difficult problems that Einstein had to solve.

THE SECOND STAGE

Einstein then came to realize the possibility of exploring, aiming at, and making use of a non-Euclidean, curved, Riemannian space. He was convinced that Newton's absolute space had had its day. This essential realization marks the second great stage of the birth of general relativity. During the summer of 1912, Gauss's theory of surfaces helped Einstein better define his program, not only because of the analogy provided by this representation—from a flat space to a curved one—but also technically, because of the curvilinear coordinates that Gauss introduced in his theory. All the more so given that (at least) two ideas suggest the need for a curved space: the revolving disk which, as we have seen, did not seem to be compatible with Euclidean geometry; and the geodesics principle, which, by generalizing the notion of a straight line would allow Einstein to retain the law of inertia in a curved space and to free the particles from any gravitational force.

Around 1820, Carl Friedrich Gauss, one of the greatest mathematicians of the nineteenth century, completed a cartographic survey of the kingdom of Hanover, a project that explains his interest in geodesy and the theory of surfaces. He then proceeded to formulate a general theory of the intrinsic geometry of surfaces. It is not surprising, therefore, to find in his treatise a definition of the length of an arc of curve on a curved surface, obtained essentially by using the Pythagorean theorem to calculate an (infinitesimal) length element from its coordinates. In a geometric space that was still that of Euclid and Newton, Gauss introduced a *metric*, a tool that defined the basic distance and made it possible to do geometry on curved surfaces. He also defined Gaussian curvature, the value of the radius of curvature at a point on a surface. These two mathematical magnitudes, the length of an arc of curve and curvature, are *intrinsic, invariant*, in the sense that they are independent of the coordinates used to calculate them.

Einstein was thus led to introduce, following Minkowski, the infinitesimal proper time line element ds. This magnitude solved several problems at once: as a geometric tool, it allowed him to calculate the *length* of arcs of curve, thus, to measure space-time, and as a physicist's tool, it defined the clock that measured proper time locally, the proper frequency of a local proper time standard. The infinitesimal proper time line element, the metric of space-time, ds, became the fundamental invariant of general relativity.

In the same way as Gauss's theory enabled the description of curved surfaces in Euclidean space, the above metric opened the way to an intrinsic description of a curved space-time. Just as the shortest lines on a surface are the geodesics, Einstein saw the material particles—the planets, the stars—which were freed from every force, travel along the extreme paths (which are, surprisingly, the slowest) in the universe: the geodesics, the paths in space-time.

As for proper time itself, *s*, it was defined only for the observers, the objects traveling along their proper trajectory, their geodesic, and it would simply be, just as in special relativity, the time indicated by the clock they carry with them, and therefore the time that defines d*s*. From a mathematical point of view, it would be enough to add up the terms, step by step, to integrate d*s* along the clock's trajectory. Proper time, *s*, was nothing but the integral of the infinitesimal element d*s* along a path in space-time. Thanks to proper time, it would be possible to measure space-time.

The structure of space-time would be determined by the distribution of matter through field equations yet to be found. It would then become possible to experiment with space-time, which was no longer given beforehand as in Newton's theory but, rather, presented as the conclusion, the solution of the problem, and which represented both the structure of space-time and gravitation; gravitation carried by space-time, cast in space-time. In this space-time, all reference frames (and not just the inertial ones) would be allowed, would be possible; from then on, all reference systems, all coordinate systems would be admissible: general covariance would therefore play a fundamental role.

But if there were no privileged coordinate system, then coordinates would no longer have any physical meaning, and that was not a simple matter. We have already seen that in special relativity only proper time has a clear meaning. It is the same thing in general relativity, where it is impossible to speak in the classical sense of the distance between two places. This is also related to the fact that the notion of rigidity is no longer meaningful. All that counts now are the actual trajectories of particles, which are nothing but the geodesics of space-time. The principle of equivalence follows from this for, if we postulate that space-time has its own structure, then the trajectories of particles, the geodesics of space-time, are completely determined, etched in space-time, and they are therefore automatically independent of the composition of the particle itself.

The universe had to be reinvented, and Einstein's principles were like tiny lamps that illuminated his way. He could make out the contours of a theory of gravitation where all the masses, all the particles, all the energy in the universe contributed to its structure: space-time was bent under the weight of matter-energy.

The equations were yet to be found. This would be the most important, the most delicate, the most confusing part of the path ahead. The broad lines were drawn, the principles instituted; but, even if they showed the way, they did not completely determine the theory, and in particular they said nothing regarding the correct field equations.

It is a very complex story, made up of breakthroughs, hesitations, mistakes, and lucky breaks. In this respect, one must emphasize the importance of the intellectual architecture comprising principles that form a conceptual network in which each one has its own particular significance and is neither absolutely necessary nor totally independent of the other principles. Some of the principles, such as Mach's, are negotiable and questionable, while others are unquestionably required, such as the principle of correspondence.

Let us now examine in greater detail the difficult birth of general relativity, of its lines of force, its guiding ideas that enabled Einstein to build—and his contemporaries to accept—that monument of modern physics in a relatively small period of time.

Throughout the years from 1912 to 1915 Einstein manipulated, tossed around, and tried to reconcile all the material he had gathered and which we have for the most part explored: his *heuristics*, his principles, the thought-experiments on which he relied, the mathematical apparatus that he was gradually mastering, and the only test available to him—the motion of the perihelion—not to mention other tests he had in mind.

During those years of relentless work, with the help of his friend Marcel Grossmann, (then professor of mathematics at the E.T.H.), supported by the interest of Besso and Ehrenfest, and spurred by the rare colleagues who also took an interest in his theories—notably Abraham (although Einstein did not think highly of his work), but also Gunnar Nordström (whose works he very much appreciated)—Einstein would translate and incorporate his ideas and principles into the mathematical structures that he was in the process of discovering, learning, and digesting. A mathematics composed of tools that were completely new for the time, fairly complex tools that had never been used before in theoretical physics. It was mostly in Ricci's and Levi-Civita's *Méthodes de calcul différentiel absolu et leurs applications*, published in French in *Mathematische Annalen*,[29] that Einstein and Grossmann would bury themselves.

The goal was to find the form of the geometric structure, the first member of the field equation. It was already clear that it would be a differential expression "of the second order" (i.e., in the second derivatives, as in Poisson's equation) involving both spatial and temporal differentials, and in which the six unknown functions of "the metric" of space-time, the gravitational potentials, were the stake. This metric, made up of a tensor of rank 2 (essentially for the same reasons the Pythagorean theorem involves squares) in ten components, would generalize that of Minkowski's space-time and would have to define the structure of space-time, that is, the infinitesimal space-time element d*s*. Einstein's strategy was necessarily twofold, for he had to couple the mathematical structure of Riemann spaces with the structures and principles of physics he wanted to put in place.

The search for the right field equations had begun, in particular that of the structure of the left-hand side of the equation, which would represent the geometry. Whose tensor would it be? Riemann's, Ricci's, Hilbert's, or Einstein's? Faced with so many possibilities, Einstein in turn hesitated, hoped, gave in, took a step forward, backed away.

It was no doubt a difficult period for him. In his personal life, the situation was not simple. While his relation with Mileva was under increasing strain, he was slowly getting involved with his cousin Elsa, who was living in Berlin: "I cannot find the time to write because I am occupied with truly great things. Day and night I rack my brains in an effort to penetrate more deeply into the things that I gradually discovered in the past two years and that represent an unprecedented advance in the fundamental problems of physics."[30] And during the same period, he wrote to his friend Heinrich Zangger:

> Do not be indignant because of my long silence! I was toiling again on the gravitation theory to the point of exhaustion, but this time with unheard-of success. That is to say that I succeeded in proving that the equations of gravitation hold for arbitrary moving reference systems, and thus that the hypothesis of the equivalence of acceleration and the gravitational field is absolutely correct in the widest sense. Now the harmony of the mutual relationships in the theory is such that I no longer have the slightest doubt about its correctness. Nature shows us only the tail of the lion. But there is no doubt in my mind that the lion belongs with it even if he cannot reveal himself to the eye all at once because of his huge dimension. We see him only the way a louse sitting upon him would.[31]

Einstein went through periods of euphoria and depression. Here is how he apologized, as he often did, in July 1913 for not having written to a friend: "I was pitiably caught up in the gravitation problem."[32] And, a few weeks later, in a letter to Lorentz, he confessed that "[his] confidence in the admissibility of the theory is still shaky,"[33] but two days later he noted that "yesterday I found out to my greatest delight that the doubts regarding the gravitation theory, which I expressed in my last letter as well as in the paper, are not appropriate."[34] In November, he was "very satisfied with the gravitational theory. The fact that the gravitation equations are not generally covariant, which still bothered me so much some time ago, has proved to be unavoidable,"[35] and, a week later, he observed: "The gravitation question has been clarified to my complete satisfaction."[36]

Throughout this period, Einstein's research would go around in circles. It is clear that Riemann's tensor had to be at the heart of the construction. It was a tensor of rank four, with four indexes, but Einstein was looking for an equation of rank two, whose origin was not hard to find: it was Ricci's tensor, obtained by "contracting" two indexes in Riemann's tensor, as Einstein did right away. But it did not seem possible to recover the corresponding Newtonian equations, which was a fundamental requirement. That prompted Einstein, for reasons that were not at all simple, to try to solve the problem by restricting the covariance, that is, by limiting, without a physical reason and in a purely ad hoc manner, the admissible coordinate systems. In addition to some unresolved technical questions, this posed a problem of interpretation: what was the meaning of this restriction of the covariance? It was a restriction that Einstein only accepted "with immense grief" and only temporarily.[37] For the next two years, Einstein wandered through all imaginable tensors and several equally unsatisfactory solutions.

In the spring of 1914, he wrote to Besso: "Now I am completely satisfied and no longer doubt the correctness of the whole system, regardless of whether the observation of the solar eclipse will succeed or not. The logic of the thing is too evident."[38] Since 1911, Freundlich had been in charge of the observation of the light-ray deflection during an eclipse, and Einstein was impatiently waiting for his results (see chapter 8).

WHO IS THE FATHER OF THE FIELD EQUATIONS: EINSTEIN OR HILBERT?

The field equations that Einstein was so anxiously looking for were not the mere translation of his principles. The principles of his theory could only help to steer his research in the direction where he was most likely to find the *right* equations. But these equations were the solution to the problem, the heart of the theory they represented; in no way were they an inevitable logical consequence of the principles. The road ahead was fraught with danger.

And of course, whoever found the field equations could rightfully claim paternity of the whole theory. For a long time after Einstein published his theory it was believed—or many *wished* to believe—that at the last moment Einstein had stumbled, that he had allowed David Hilbert, one of the greatest mathematicians in the world, to overtake him. The question was asked, could Einstein have "borrowed" the missing term in the field equations? It was a question that long remained without an answer (a postcard from Hilbert to Einstein is missing)—with the presumed "borrowing" casting a painful doubt—but nevertheless Einstein was awarded most of the credit as father of the theory.[39]

In short, it was thought that Hilbert had "got out" the general relativity field equations one week—five days, to be precise—before Einstein.[40] A scoop! Imagine, Einstein's equations would really have been Hilbert's! A good reason to render Einstein's results relative—pun intended. Never mind the vast amount of work he had accomplished throughout 1907–15 (as we have seen in the preceding chapters)—the principles he defended, the connections he created, and the physical interpretation he meticulously built. All that really mattered were the field equations, this high-precision mechanism that completed the theory and made the calculations possible; calculations thanks to which physical facts were accounted for and predicted. Those equations represented the summit of the theory, which no one has revisited since, except Einstein himself in 1917 (see chapter 15).

In the summer of 1915, Einstein visited Göttingen, where David Hilbert had recently demonstrated an interest in gravitation. At the end of June and beginning of July, Einstein gave a series of six talks in Göttingen, where he "had the great pleasure of seeing that everything was understood down to the details."[41] During the talks he "was able to convince Hilbert of the general theory of relativity."[42] Hilbert, of whom Einstein was said to be "quite enchanted"[43] was "a man," he wrote to Zangger, "of astonishing energy and independence in all things."[44] He spoke in those terms of Hilbert, a man whose mathematical knowledge was vast compared to that painfully acquired by Einstein and who was quite capable of overtaking him at the very last moment.

November was a crucial month. Einstein published no fewer than four articles, while he corresponded nonstop with Hilbert, exchanging papers and ideas. On 15 November, he received a letter, to which he replied: "Your analysis interests me tremendously, especially since I often racked my brains to construct a bridge between gravitation and electromagnetism."[45] He asked Hilbert for a copy of his article but declined an invitation to go to Göttingen. Suddenly, on 18 November, to be precise, after

Figure 5.2. David Hilbert. (© AIP Emilio Segrè Visual Archives)

receiving and reading the first draft of an article that Hilbert would soon publish, the tone changed. An annoyed Einstein began a letter to Hilbert by denying the originality of his approach ("exactly what I found these past weeks"). Then, not without bitterness, he added:

> The difficulty was not in finding generally covariant equations for the $g^{\mu\nu}$'s; for this is easily achieved with the aid of Riemann's tensor. Rather, it was hard to recognize that these equations are a generalization, that is, a simple and natural generalization of Newton's law.[46]

Finally, Einstein told him that he had the very same day submitted to the Academy

> a paper in which I derive quantitatively out of general relativity, without any additional hypothesis, the perihelion motion of Mercury discovered by Le Verrier. No gravitation theory had achieved this until now.[47]

In chapter 7 we will come back to this first fundamental success. Einstein had just succeeded in accounting for (without any ad hoc hypothesis!) Mercury's anomaly that had so troubled astronomers for more than half a century. He did not yet have the final equations of the theory but only those that, being valid in vacuum, were enough for this particular calculation. He no doubt returned to work with renewed eagerness the following week. Wasn't he finally on the right track, since he had just found the correct motion of the perihelion? But didn't he fear to be beaten at the wire by Hilbert, who may announce the final field equations before him?

A week later, on the 25th, he submitted his article on the field equations "with second member," the very same equations that Hilbert seemed to have had accepted for publication on the 20th. There is a *diagonal* term missing in Einstein's equations, an essential term that explicitly appears in Hilbert's article, which was published at the end of March of the following year.[48] That is the reason why many authors, in their comments, refer to Einstein's equations as "Einstein-Hilbert." But how is it then possible that, in a note on the first page of his article, Hilbert refers to all the November articles and in particular to Einstein's final paper, the one submitted on the 25th? And in the body of the article Hilbert observes that his differential equations are "in agreement with the magnificent theory of general relativity established by Einstein in his later papers."[49] Moreover, it turns out that in Hilbert's original manuscript, which has recently been found, the final field equations are absent.[50] In fact, Hilbert had made some changes to his paper after the acceptance date of 20 November but, unfortunately, this is the date that appears in the published article, while the date of the revision is not mentioned. In the published version, Hilbert gives the final field equations, for which the knowledge of Einstein's results appears to have been crucial, as he acknowledges in the definitive version of the article finally published on 31 March 1916.

The "right" tensor, the one which is needed in the left member, is now called "Einstein's tensor"—the least that one would expect. In fact, his tensor, $E_{\mu\nu}$, is extremely similar to Ricci's tensor $R_{\mu\nu}$, and coincides with it in the case of the vacuum. Not only did Einstein not obtain it from Hilbert but, three years earlier, while he was working with Grossmann, they had considered Ricci's tensor only to later abandon it because they thought that with it they could neither recover Newton's equations nor satisfy any law of conservation of energy.

General relativity was thus born on 25 November 1915 in an article that appeared in the *Proceedings of the Royal Academy of Sciences of Prussia*, where Einstein had published since he joined the Berlin Academy. The field equations take the now classical form:

$$E_{\mu\nu} = R_{\mu\nu} - 1/2\, Rg_{\mu\nu} = \chi T_{\mu\nu}.$$ [51]

On 26 November, Einstein wrote to his friend Zangger that "[t]he theory is beautiful beyond comparison," but he could not resist adding: "However, only one colleague has really understood it, and he is seeking to "partake" [*nostrofieren*][52] in it (Abraham's expression) in a clever way. In my personal experience I have hardly come to know the wretchedness of mankind better than as a result of this theory and everything connected to it. But it does not bother me."[53]

He was, of course, talking about Hilbert, a person he had so liked only recently and who, a month later, would send him a letter of congratulation on his appointment as correspondent member of the Göttingen Royal Society of Science. Einstein, in a generous mood—isn't he elated by his success?—replied to Hilbert politely but not without sincerity:

> Highly esteemed colleague,
> I thank you for your kind communication regarding my election as corr. member. On this occasion I feel compelled to say something else to you that is of much more importance to me. There has been a certain ill-feeling between us, the cause of which I do not want to analyze. I have struggled against the feeling of bitterness attached to it, and this with complete success. I think of you again with unmixed geniality and ask you to try to do the same with me. Objectively it is a shame when two real fellows who have extricated themselves somewhat from this shabby world do not afford each other mutual pleasure.
>
> Best regards, yours
> A. Einstein[54]

As soon as the final article was published, he apologized to Sommerfeld for not having replied to his earlier letter sooner: "But in the last month I had one of the most stimulating, exhausting times of my life, indeed also one of the most successful. I could not think of writing."[55] And, after having described in great detail his difficulties with the covariance and the tensors, he returned to the perihelion: "The splendid discovery I then made was that not only Newton's theory resulted in first order approximation, but also Mercury's perihelion motion (43" [seconds of arc] per century) in second order approximation." And he added a bitter remark aimed at Hermann Struve, director of the Royal Observatory of Prussia: "Freundlich has a method of measuring light deflection by Jupiter. Only the intrigues of pitiful persons prevent this last important test of the theory from being carried out."[56]

A few days later, he sent Sommerfeld his published articles, and he added: "Be sure to have a look at them; it is the most valuable finding I have made in my life."[57] In January 1916, he would write to his friend Ehrenfest: "Imagine my delight at realizing that general covariance was feasible and at finding out that the equations yield Mercury's perihelion motion correctly. I was beside myself with joy and excitement for days."[58] In the first paper of the series that led him toward general relativity, in November 1915, Einstein already could not resist declaring his own amazement: "Scarcely anyone who has understood this theory can escape from its magic."[59]

"Who has understood this theory . . ." Alas, the physicists who would be unmoved by this magic, by this very real beauty, would be many, and for quite a long time. The theory was complete; it now had to be "verified," properly understood, and accepted. In chapters 10 and 11 we will examine in detail the obstacles to understanding and acceptance of general relativity; it was a strange phenomenon and one of the dark chapters of theoretical physics in the first half of the twentieth century.

Einstein now had at his disposal the new field equations, whose solution was a curved *space-time* representing gravitation. How do these equations work? We shall answer that in several steps. First, we must state a gravitation problem; on the right side, we need a distribution of matter, "T," whose gravitational field we wish to know. Einstein's field equations have two sides, as all equations do—two faces, so to speak: $E_{\mu\nu} = \chi T_{\mu\nu}$. The right term, "$T_{\mu\nu}$" represents the matter at the origin of the field. This material distribution will affect the left term of the equation, "$E_{\mu\nu}$" Einstein's tensor, whose discovery we have just briefly discussed and which contains the mathematical tools to forge, to express, the geometry of the solution space-time. Solving these equations then consists in finding the space-time that is the solution of the problem—a Riemannian space-time.

In the case where the gravitational field has no source, or, equivalently, if the term in "T" is identically zero, indicating that there is no mass anywhere, then, since no gravitation manifests itself, the "solution-space" of the problem must be "flat" and, being a space-time, it must necessarily be that of special relativity—Minkowski's. In a flash, we are back at the starting point.

But, in general, we will have to specify how the bodies (i.e., the test particles, which are so small that they do not contribute to the original field) move in a curved space-time. What is the equation of their trajectory? To this question, we have an immediate answer: the particles traveling in this space will have no choice; they will have to follow the geodesics, the extreme paths in space-time. Therefore, once the geometry of space-time is known, its possible trajectories—all possible trajectories—follow.

However, there is not just one space-time that is a solution of Einstein's field equations. All depends on what we put "on the right." In the following months and years, there would be a search for solutions to some simple questions, first of all: what is the space-time corresponding to the gravitational field of a spherical star such as the Sun? Karl Schwarzschild, a German astrophysicist, would be the first to solve the equations exactly. But the interpretation of this solution was far from obvious (we shall return to the topic in chapter 12), and it would only be satisfactorily understood fifty years later. The black holes that result in certain situations would not be imagined until the end of the 1960s (see chapters 13 and 14).

NOTES

1. Brian, 1996, p. 31 and p. 72.
2. Minkowski to Einstein, 9 October 1907, *CPAE*, vol. 5, p. 77.
3. An outline of which appears in Poincaré, 1906, p. 168.
4. Einstein, 1907, *CPAE*, vol. 2, pp. 433–488.
5. Einstein, 1905b.
6. Einstein, 1907, German original in *CPAE*, vol. 2, p. 464.
7. Einstein, 1907, German original in *CPAE*, vol. 2, p. 465; English translation vol. 2, p. 288.
8. Einstein, 1907, German original in *CPAE*, vol. 5, p. 476.
9. Einstein, 1907, German original in *CPAE*, vol. 2, p.480; English translation vol. 2, p. 307. Einstein employ here the term gravitational "potential." He could have just as well spoken of gravitational "field." Curvature is the only concept correctly defined—in an intrinsic way, although it is not an immediately measurable physical magnitude. Hence, the need for terms such as "intensity of the gravitational field" and "gravitational potential," which are not well defined but nonetheless suggestive. These terms come from Newton's and Maxwell's theories, and in that context they are rigorously defined.

10. Einstein, 1907, German original in *CPAE*, vol. 2, p. 481; English translation vol. 2, p. 307.

11. Einstein to C. Habicht, 24 December 1907, German original in *CPAE*, vol. 5, p. 82.

12. Brian, 1996, p. 70.

13. Brian, 1996, p. 74. Quoted from Pais, 1982, pp. 185–186.

14. A. Sommerfeld to H. A. Lorentz, 26 December 1907, Rijksarchief in Noord-Holland, Haarlem, H. A. Lorentz Archives.

15. In connection with this see Feuer, 1982, p. 355, and the Einstein-Born correspondence, Born, 1969, pp. 164–172.

16. M. Planck, 1910, quoted from *CPAE*, Introduction, vol. 5, p. xxxvi.

17. Cf. Stachel, 1980.

18. Einstein, 1911, German original in *CPAE*, vol. 3, p. 486; English translation in Einstein, 1923, p. 99.

19. Ibid.

20. Einstein, 1911, German original in *CPAE*, vol. 3, p. 496.

21. Einstein, 1911, German original in *CPAE*, vol. 3, p. 496 and Einstein, 1923, p. 108. Einstein was interested in Jupiter because it is the largest planet in the Solar system. But, as he observed, the effect is much too weak and was never measured despite E. Freundlich's efforts.

22. In this respect, see "Editorial Note: Einstein on Gravitation and Relativity: the Static Field", *CPAE*, vol. 4, pp. 122–128.

23. Einstein to M. Besso, 26 March 1912, German original in *CPAE*, vol. 5, p. 436–437; English translation, vol. 5, p. 278.

24. Einstein to M. Besso, 1 January 1914, German original in *CPAE*, vol. 5, p. 588; English translation, vol. 5, p. 374.

25. Einstein to A. Sommerfeld, 29 October 1912, *CPAE*, vol. 5, p. 505; English translation, vol. 5, p. 324.

26. Einstein to Ludwig Hopf, 20 February 1912, *CPAE*, vol. 5, p. 418; English translation, vol. 5, pp. 266–267.

27. Einstein to P. Ehrenfest, 28 May 1913, *CPAE*, vol. 5, p. 523; English translation, vol. 5, pp. 334–335.

28. Mach, 1883.

29. Ricci and Levi-Civita, 1901.

30. Einstein to E. Löwenthal, February 1914, *CPAE*, vol. 5, pp. 597–598; English translation, vol. 5, pp. 378–379.

31. Einstein to H. Zangger, 10 March 1914, *CPAE*, vol. 5, pp. 601–602; English translation, vol. 5, pp. 380–381.

32. Einstein to J. Laub, 22 July 1913, *CPAE*, vol. 5, p. 538; English translation, vol. 5, pp. 343–344.

33. Einstein to H. A. Lorentz, 14 August 1913, *CPAE*, vol. 5, p. 547; English translation, vol. 5, pp. 349–351.

34. Einstein to H. A. Lorentz, 16 August 1913, *CPAE*, vol. 5, p. 552; English translation, vol. 5, pp. 352–353.

35. Einstein to L. Hopf, 2 November 1913, *CPAE*, vol. 5, pp. 562–563; English translation, vol. 5, pp. 358–359.

36. Einstein to P. Ehrenfest, 7 November 1913, *CPAE*, vol. 5, p. 563; English translation, vol. 5, p. 359.

37. It is a feeling that Einstein only fully acknowledged after the question was resolved. Einstein to D. Hilbert, 18 November 1915, German original in *CPAE*, vol. 8, p. 201; English translation, vol. 8, p. 148.

38. Einstein to M. Besso, March 1914, *CPAE*, vol. 5, p. 604; English translation, vol. 5, pp. 381–382.

39. In this connection, cf. Corry et al., 1997.

40. For example, Fölsing, 1997, quoted by Corry et al., 1997.

41. Einstein to A. Sommerfeld, 15 July 1915, *CPAE*, vol. 8, p. 147; English translation, vol. 8, p. 111.

42. Einstein to H. Zangger, July-August 1915, *CPAE*, vol. 8, p. 154; English translation, vol. 8, p. 116.

43. Einstein to A. Sommerfeld, 15 July 1915, *CPAE*, vol. 8, p. 147; English translation, vol. 8, p. 111.

44. Einstein to H. Zangger, July-August 1915, *CPAE*, vol. 8, p. 154; English translation, vol. 8, p. 116.

45. Einstein to D. Hilbert, 15 November 1915, *CPAE*, vol. 8, p. 199; English translation, vol. 8, p. 146–147.

46. Einstein to D. Hilbert, 18 November 1915, German original in *CPAE*, vol. 8, pp. 201–202; English translation, vol. 8, p. 148.

47. Ibid.

48. In a letter to H. Weyl, Einstein had some harsh words for Hilbert's article: "Hilbert's assumptions about matter appear childish to me, in the sense of a child who does not know any of the tricks of the world outside." Einstein to H. Weyl, 23 November 1916, *CPAE*, vol. 8A, p. 367.

49. Quoted by Corry et al., 1997, p. 1273.

50. Ibid.

51. In the modern convention used here, Greek subindexes range from 1 to 4, the first three representing the spatial components (sometimes, they range from 0 to 3, with 0 playing the role of time). Notice that the fact of assigning a particular digit to time suggests a kind of refusal to accept a perfect symmetry between the roles of time and space. We shall see an example of this in Kruskal's representation (chapter 13).

52. In fact, Einstein employs here a neologism (*nostrofieren*) that can only be translated approximately as "partake."

53. Einstein to H. Zangger, 26 November 1915, *CPAE*, vol. 8, p. 205; English translation, vol. 8, pp. 150–151.

54. Einstein to D. Hilbert, 20 December 1915, *CPAE*, vol. 8, p. 222; English translation, vol. 8, p. 162–163.

55. Einstein to A. Sommerfeld, 28 November 1915, German original in *CPAE*, vol. 8, p. 206.

56. Ibid., p. 208; English translation, vol. 8, pp. 152–153. Einstein paved the way to the measuring of this effect of light deflection by the Sun's and Jupiter's gravitational fields in his 1911 article.

57. Einstein to A. Sommerfeld, 9 December 1915, German original in *CPAE*, vol. 8, p. 217.

58. Einstein to P. Ehrenfest, 17 January 1916, *CPAE*, vol. 8, p. 244; English translation, vol. 8, pp. 177–179.

59. Einstein, 1915, p. 779, *CPAE*, vol. 6, p. 216.

CHAPTER SIX

General Relativity: A Physical Geometry

EINSTEIN'S INTEREST in philosophy, the success of his special theory, and later the construction of general relativity led him to ponder the meaning of a physical theory and the question of its truth, and to wonder about the differences and connections between mathematics and the physical sciences. He had plenty of reasons to do so.

"Science" is often considered to encompass mathematics (first of all, of course), then physics, followed by the other physical and natural sciences, more or less in order of importance of their relation to mathematics. But hardly any distinction is made between the methods, the status, and the results of those sciences, mathematics in particular. What is the place of mathematics? Is it only a tool of the physical and natural sciences? Should we speak of mathematical sciences? Shouldn't we clearly distinguish between mathematics and the natural sciences? What is the essence of mathematics? What is the meaning of "nature" for a mathematician?

Mathematics is a science like no other; it is not a natural or physical science. Its status is totally different from that of physics or the natural sciences because it has no direct connection with experience. In mathematics there are no experiments of the type that can be performed in physics.[1] This is no small difference. Mathematical propositions are certain and indisputable, whereas those of all the other sciences are up to a certain point questionable and, as Einstein put it, are "in constant danger of being overthrown by new discovered facts."[2]

This essential point, this difference of structure, of status, of relation to truth, is clarified by general relativity in a very simple and convincing manner through the question: is geometry part of mathematics or part of physics? A question that may appear absurd, and yet . . .

Let us take a general look at the geometry of the universe. Well after Euclid, in Descartes' and Newton's time, the *practical* geometry employed by the philosophers of nature was indistinguishable from Euclidean geometry. It was given a priori as the fundamental shape of space. We now know that the problem is not that simple and that the geometry of the universe is Euclidean only in the first approximation. After all, light rays do not travel in straight lines but follow well-defined trajectories in the geometry of the universe, very special geodesics. The question then arises: what is the geometric structure of the universe? This is the central question in general relativity, and one that suggests the possibility of geometric experiments. The deflection of light rays by a star's gravitational field is an experimental fact, something that can be observed, and which raises the question of the geometry of the universe. Would it be possible that, contrary to what I have stated earlier, geometry could be subjected to experimental verification? Yes and no: we must distinguish between practical and theoretical geometry and establish a science of space of which general relativity would be the first real theory, the practical geometry. It is precisely this distinction, one that mathematicians did not see or make until the nineteenth century, that general relativity imposes.

WHEN IS A LINE A STRAIGHT LINE?

Let us go back in time to our high school years, to plane geometry class, Pythagoras's theorem, and Thales' theorem (on parallels) and begin with an infinitely simple question regarding straight lines. So simple, the best students might say, perhaps a little too simple—as we are going to see.

Euclid's fifth postulate states that "through a point not on a given line passes one and only one parallel to that line."[3] The birth of non-Euclidean geometries in the mid-nineteenth century challenged the absolute *truth* of Euclid's geometry and raised the question that we, as physicists, ask ourselves: "Is the universe, the universe that is given to us, Euclidean?" The word "geometry" comes from the Greek *geometria*, from *ge*, "Earth," and *metron*, "measure." It was then about "measuring the Earth," that is, land surveying. In those times, the rise in level of the Nile's waters resulted in floods that forced ancient geometers to redefine the land plots every year. It is believed that geometry had its roots there, and yet, no effort was spared, especially in the seventeenth and eighteenth centuries, to conceal and forget the secular origins of the queen of sciences, origins to which general relativity returns us. Let us see precisely how.

If we try to define "straight line" in Euclidean geometry, we realize that our attempt falls short of a formal definition. In his *Elements*, Euclid said, "A straight line is one that lies evenly with respect to the points on itself."[4] It is an ambiguous definition involving optical intuition. Efforts were made to eliminate physical reality, to cast a mathematical definition as truly and completely axiomatic and independent of any image of reality, a definition that would be in some sense eternal. At the same time, classical science was being built on what it believed to be a solid rock: Euclidean space, Cartesian space, and Newton's absolute space; in short, the space of classical geometry, of the Pythagorean theorem of our youth.

A reconstruction of geometry independent of—but also at the expense of—reality had to wait until the nineteenth century, when the parallel axiom was called into question. A straight line was then defined by its properties, those of superposition and displacement (by sliding). No previous knowledge or intuition of this mathematical object was assumed, except implicitly; only the validity of the axioms defining it. It was a conceptual scheme from which all practical content was removed. But then, how to construct a straight line from this axiomatic definition? In fact, no mathematical definition that is both simple and precise is available. There is no longer a connection between the reality of the straight line and its conceptual definition. In a word, the question is not as obvious as it was thought to be in classical times.

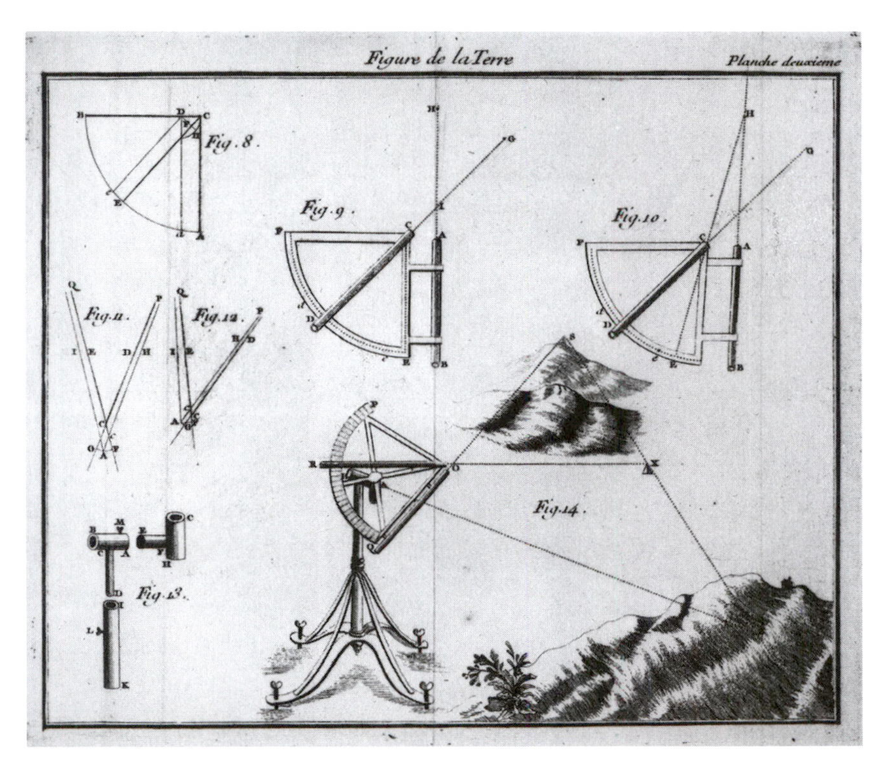

Figure 6.1. Straight line and light ray. *La figure de la Terre*, Pierre Bouguer, 1749. (Photo Observatoire de Paris)

We must therefore distinguish, as Einstein did, between *practical* and *axiomatic* geometry. A distinction that once again was first made by Poincaré in the chapter "Space and Geometry" of his monumental treatise *Science and Hypothesis*. Poincaré opens that door when he distinguishes between "geometric space," and "representation space," while he accepts that "someone educated in an environment where these laws . . . would be upset could have a geometry very different from ours."[5] But in his conclusion he unambiguously expressed his ambivalence: "We can see that experience plays an indispensable role in the genesis of geometry; but it would be a mistake to conclude that geometry is an experimental science, even in part."[6]

A few paragraphs later in the conclusion, he observes that "experience guides us in that choice but it does not impose it on us; it shows us not which geometry is the truest, but which one is the most convenient."[7]

If there were a choice, no one doubts that physicists would happily make it. The curved, non-Euclidean space was, and still is today, an obligatory choice; one which despite its obvious and extraordinary success stands in the way of a global description of the physical world, due to the effect on the shaping of our concepts of two thousand years of mathematics cast in the language of Euclid. Nevertheless, in order to understand Poincaré's approach, physics is not enough; psychology and his attachment to the structure of science at the time also play a role; the usual conflict of generations.[8]

How do we construct a straight line in practical geometry? Ruler and compass immediately come to mind. These are the tools we generally use to draw figures. A ruler? Would that solve the problem? A ruler (i.e., a straight line) to define a straight line? We are going around in circles, and the problem remains. In Euclidean geometry, often without mentioning it, we are *given* the lines and planes and the postulates that go with them. We are back to square one.

From a more practical point of view, let us look at the way engineers, architects, bricklayers, and astronomers proceed. What kind of ruler do they use? How do they guarantee that the wall they build is "flat," that it is a plane? They may employ a ruler, made to that effect, to "smooth out" their wall. But how do they manufacture that ruler? Or, more precisely, how do they verify that their wooden ruler is always *straight*, despite the humidity, or that it has not been twisted by the heat, if it is a metal one?

The bricklayer or the architect will take a sight; he will close one eye and, if he can make all the points of the ruler or wall to line up, then he is satisfied that the ruler is straight, that the wall is a plane. That is also what astronomers do with an instrument called, appropriately, a sight, and what surveyors also do with a goniometer, in order to draw their maps. They make the mark on the graduated circle of their telescope coincide with their eye and the star—or the place whose position they wish to determine. In a nutshell, it is through a sight that a straight line is most accurately defined. A sight, that is, the trajectory of a light ray.

We must once again return to Poincaré. In *Science and Hypothesis*, he does not hesitate to take a first step in accepting that in astronomy a straight line is "simply the trajectory of a light ray." But he refuses to take a second step that would take him to a non-Euclidean geometry, for he basically says that, even if observation would warrant it, we should not abandon Euclidean geometry but rather "acknowledge that light does not propagate strictly in a straight line." Needless to say," he adds, "everybody would consider this to be the most advantageous solution," and he concludes: "Euclidean geometry has therefore nothing to fear from new observations."[9] On reading him, we may better understand the wall of silence he built around himself at the end of his life as an expression of his disagreement with the work then being carried out by Einstein.

In the textbooks on perspective and practical geometry used by surveyors, a straight line is simply a light ray. Granted, you would say, that is true in practical geometry, but don't our surveyor mathematicians have another tool in their toolkit? They don't, really. In fact, they postulate the existence of mathematical objects having all the properties of straight lines without ever wishing to construct them. They work with the axioms defining those objects. At some time or other, if they are teaching, or to better see what they are doing, they may resort to simple drawings (necessarily on a plane) or use a ruler. But of course nothing of this appears in their proofs.

Conversely, it is clear that Euclidean geometry is the result of a reformulation aimed at removing all reference to reality and in particular to sight. In the post-Euclid centuries, the fields of practical geometry (optics, in fact) and axiomatic geometry (mathematics) have become practically disjoint. There is a definite intention, and a perfectly justifiable one, to work with pure concepts that are independent of reality, and this separation is at the very

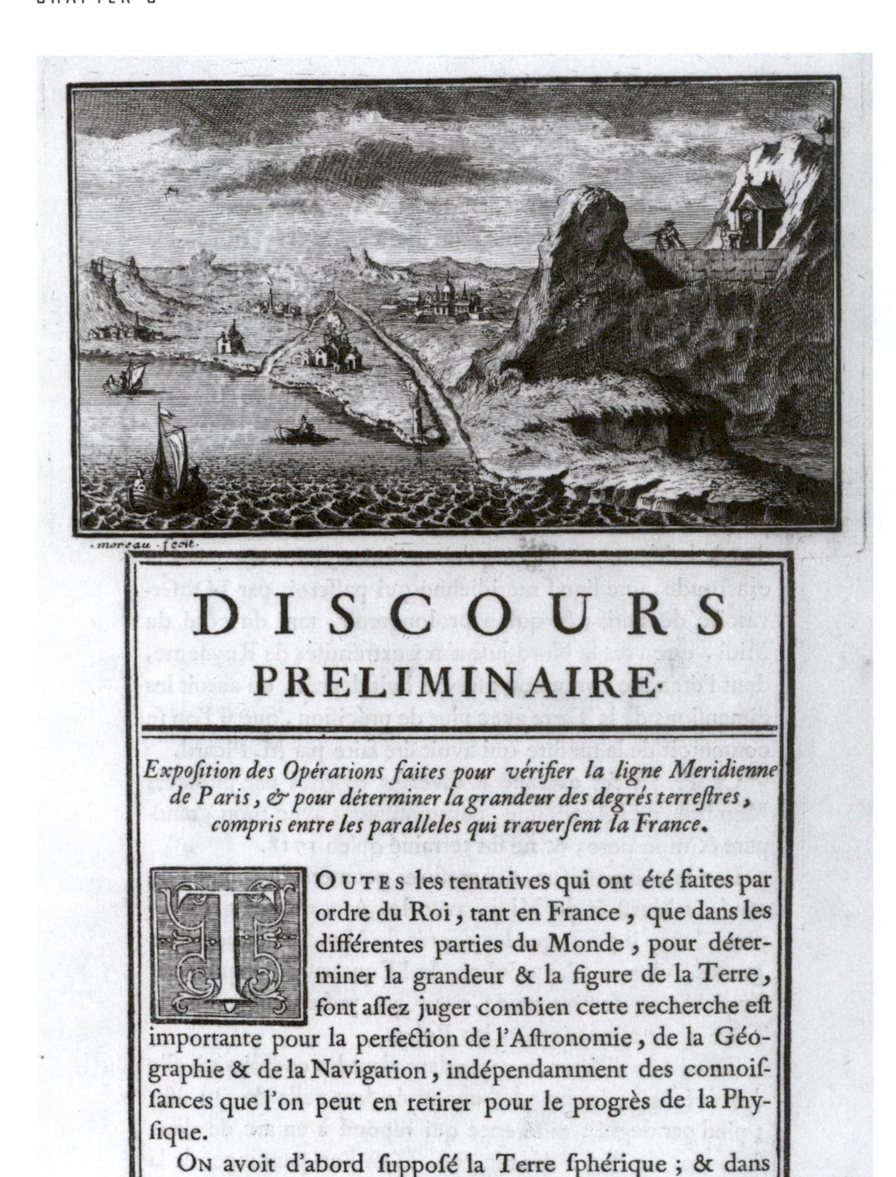

moreau · fecit.

DISCOURS PRELIMINAIRE.

Expoſition des Opérations faites pour vérifier la ligne Meridienne de Paris, & pour déterminer la grandeur des degrés terreſtres, compris entre les paralleles qui traverſent la France.

TOUTES les tentatives qui ont été faites par ordre du Roi, tant en France, que dans les différentes parties du Monde, pour déterminer la grandeur & la figure de la Terre, font aſſez juger combien cette recherche eſt importante pour la perfection de l'Aſtronomie, de la Géographie & de la Navigation, indépendamment des connoiſſances que l'on peut en retirer pour le progrès de la Phyſique.

ON avoit d'abord ſuppoſé la Terre ſphérique ; & dans

A

Figure 6.2. Light structures space. *The meridian of the Observatoire de Paris, verified through new operations.* Cassini de Thury, 1744. (Photo Observatoire de Paris)

basis of the existence of mathematics. Reality is deliberately banned from mathematical thought, so much so that, in the seventeenth century, "geometry" and—even more so—"geometer" were taken to mean "mathematics" and "mathematician."

Mathematics is therefore a science like no other. After the geometers, the mathematicians work on conceptual schemes void as much as possible of all reality (except an archaic one) and all they need is a pencil. If a computer happens to be used in a proof, this is a source of concern. Their objects are virtual ones, products of the mind, and their work is only meaningful if it is independent of all material reality. The methods and axioms of mathematics do not refer to real objects but to "objects of our mere imagination," as Einstein put it,[10] from which follows that mathematical propositions "are absolutely certain and indisputable"[11]; they are indisputable and irrefutable. Mathematical theorems cannot be experimentally verified. Proofs are formal and based essentially on definitions by convention and on methods that are proper to the discipline. Mathematical theorems are therefore eternal. Such is the case, for instance, of Pythagoras's theorem, which has not changed in more than two millennia. This is why there is no drama as the history of mathematics unfolds; its theories and propositions are immortal.

In that sense, mathematics is totally different from physics and the natural sciences, where everything begins and ends with reality. The connection between those sciences and experience, between them and the observation of reality, is fundamental. And it is precisely for this reason that physical theories are mortal. A physical theory cannot be *true*, it can only be *right* or false, as in Newton's case; at a certain point in history, his theory will be outdated. Natural sciences can only represent nature, they do not pretend to be true in a strong sense.

To sum up, mathematics as a science is a kind of art that physics (to mention only that science) uses as a tool, one that fits it sometimes very well, like a glove, and not without reason for both have a partly common origin: a geometry embedded in reality.

We have now some elements to tackle the question of the "unreasonable effectiveness of mathematics,"[12] to try to understand why mathematics applies so well to the objects of reality. It is not necessarily because reality is Platonic, or because geometry (Euclidean, of course) is at the heart of the physical universe; it is also because the physicist-geometer lives at the heart of geometry, of reality. In short, the notion of this unreal mathematics floating above a purely imaginary universe raises many questions, interesting questions we have not finished asking—not to mention answering.

PRACTICAL GEOMETRY AND AXIOMATIC GEOMETRY

There are, then, two kinds of geometry. The geometry of our high school days is the cradle of a geometric practice that we will call axiomatic geometry to distinguish it from the practical geometry of the Egyptians of the Nile and also of our astronomers.

To be sure, the almost perfect Euclidean character of space (the fact that our physical, terrestrial space is essentially flat) precludes any possible conflict. The light ray that represents a straight line is in fact straight, for space is (almost) not curved. Axiomatic geometry is the expression of practical geometry, and practical geometry is a physical science whose theory is precisely axiomatic geometry.

In simple terms, since antiquity we have constructed a geometry adapted to the space we live in and which describes it. This is hardly surprising, and it is consistent with history and experience (and psychology!). On the one hand, we have a practical geometry, whose instrument is the light ray that plays the role of a ruler. On the other, we have an axiomatic geometry that separated itself as much as it could from practical geometry and that has long refused any geometry other than the Euclidean. Axiomatic geometry has presented itself as Euclidean, while practical geometry was rejected as part of technology. Euclidean geometry became a dogma. The whole of mathematics was largely built by analogy with geometry, and science itself constructed its logic from the framework offered by geometry. Everything was geometrized. Even algebra had trouble ridding itself from the influence of geometry, the queen of sciences. Kant symbolized the dogma of the Euclidean nature of space by postulating its a priori character. The principles of Euclidean geometry were then supposed to preexist experience. It was even thought that our minds were so made that we could not see the world other than as Euclidean; in Kantian terms: an a priori synthetic truth.

Surely, in the mid-nineteenth century this dogma gradually collapsed under the pressure of the discovery, independent of any experience, of non-Euclidean geometries. Axiomatic geometry then opened itself up to other geometries, which no longer accepted the parallel dogma. But a physical science of space, a science of practical geometry was still missing. Space was always taken for granted and it was Euclidean. And why not, since it was still possible to declare at the beginning of the twentieth century that, to a high degree of accuracy, space was in fact Euclidean. It would take the relativistic revolution to call into question the dogma of absolute space. A revolution that began with special relativity and the introduction of space-time and continued with the birth of a real theory of space—general relativity—whose validity would be reinforced by the observation of the deflection of light rays (see chapter 8), an experiment that, more than any other, showed that the universe is not Euclidean.

At the present time, the only theory of space, the only theory that studies the question of the true structure of the universe as it is (and as it evolves) is general relativity. With general relativity, Einstein created a new science: the science of space, of space-time. It is as simple as that. But it had still to be accepted.

NOTES

1. Only experiments of another kind, internal (to the mathematician), archaic, and far from being accepted by all mathematicians. There are also nontraditional computer simulations and experiments. I will not dwell here on the status of mathematics or mathematical experiments, which are difficult and controversial questions. On this topic see Connes et al., 2000, and Postel-Vinay, 2000.

2. Einstein, 1921, p. 123; English translation in Einstein, 1954, p. 232.

3. See Heath, 1956.

4. Ibid.

5. Poincaré, 1902, p. 84.

6. Ibid., p. 90.

7. Ibid., p. 91.

8. In this connection, cf. Feuer, 1982, pp. 61–66.

9. Poincaré, 1902, p. 93.

10. Einstein, 1921b, p. 123; English translation in Einstein, 1954, p. 232.

11. Ibid.

12. From the title of a famous article by Eugene Wigner (Wigner, 1960).

CHAPTER SEVEN

Relativity *Verified*: Mercury's Anomaly

IN NOVEMBER 1915, Einstein finally had at his disposal a relativistic theory of gravitation that had taken him almost eight years to construct. Now it remained to put it to the test. This was certainly not just a formality. Having a conceptually satisfactory theory was not enough; the theory had to account for observed facts. But which facts? Back in 1907, he had devised three classical tests that for almost fifty years would be the only tests of his theory; three tests that could tell apart the new theory of gravitation from that of Newton's: an anomaly that the latter could not explain and two novel effects that it did not predict. Einstein's gravitation theory hung on those tenuous effects for almost half a century.

The first observed fact concerned the motion of the perihelion of the planets, Mercury's in particular, which since the mid-nineteenth century, Newton's theory had been unable to explain in a satisfactory way. Could the new theory account for Mercury's anomaly? The theory's credibility was at stake, as Einstein was well aware. Early on, he deduced from his theory the advance of Mercury's perihelion, an essential first step that reassured him and helped to shield the theory against the criticism that inevitably followed.

A second, completely new consequence: the prediction of an effect of gravitation on the propagation of light rays. Light propagates in a straight line only when there is no accumulation of matter in its vicinity. If a light ray travels near a massive body, it will be subject to the latter's gravitational field and deflected. In chapter 8, we shall examine in detail the delicate verification of this effect, the theory's second test, which required the extremely precise observation of solar eclipses. This took place after the Great War, in June 1919.

A third test was planned: to understand and measure the influence of gravitation on the working of clocks situated in different locations and subject to different gravitational fields. Those clocks, whose rate can be monitored, are everywhere in the universe: they are the spectral lines of atoms, and, in a gravitational field, these lines should be red-shifted.

In a certain sense, this third test, also known as the Einstein effect, was more crucial than the first two, for it was a direct consequence of the principle of equivalence and did not require the whole apparatus of Einstein's theory. It was, rather, a test of one of the principles of the theory. Many observations and experiments were necessary before its validity was fully accepted in 1960, when an experiment on Earth was set up to put it to the test. This was the first *experiment* in general relativity (see chapter 9).

In all three tests, the effects in question are very small, very weak. Quantitatively, they are of "$2GM/rc^2$" order, a term that "calibrates" the action of gravitation and is proportional to the mass M creating the gravitational field and inversely proportional to the distance, r, between M and the object subject to the field (see box 5). The smallness of this term explains the enormous difficulties involved in testing Einstein's theory.

FROM URANUS TO VULCAN

At the beginning of the seventeenth century, Kepler showed, thanks to Tycho Brahe's observations, the elliptical character of Mars' orbit. One hundred and fifty years later, Newton's theory of gravitation stipulated that the planets had to obey the laws formulated by Kepler, as long as they were not perturbed by the presence of other planets close by. Eighteenth- and nineteenth-century geometers and astronomers carried out an exhaustive analysis of these perturbations, which resulted in the determination of the actual motion of the planets. The Moon posed a difficult problem for Alexis-Claude Clairaut who, being unable to account for its motion, was for a while tempted to abandon

Box 5. $2GM/rc^2$

Let us take this opportunity to mention that $2GM/rc^2$ is a fundamental expression in relativistic gravitation, characteristic of the intensity of the gravitational field and which we often encounter in general relativity, as for example in the term giving the advance of the perihelion.

$2GM/rc^2$ is an indication of the intensity of the field of a pointlike or almost pointlike body of mass M situated at a distance r, where G is the gravitational constant and c the speed of light. $2GM/rc^2$ is a dimensionless term, whose intensity is very small with respect to 1.

On the Earth's surface, it is of order 10^{-9}, and of 10^{-6} on that of the Sun. On the surface of a white dwarf, whose density is much greater than the Earth's, it is already of order 10^{-4}. But on the surface of a neutron star, a star infinitely denser than a white dwarf, its order may reach 10^{-1} and even 0.4. If the order reaches 1, very strange things happen: space-time closes up because the star has collapsed within its horizon—a black hole is created (see chapters 12 to 14).

Newton's $1/r^2$ law by adding a term in $1/r^4$. But he finally succeeded in solving the problem and Newton's system remained intact. In the mid-nineteenth century, the astronomers, mathematicians, and celestial mechanics were optimistic. Within the framework of Newton's theory of gravitation, and thanks to their skill, they had managed to account for almost all the planetary motions and were confident that they would be able to explain the remaining ones.

The most spectacular moment of these developments was the discovery of Neptune in 1846.[1] Uranus, discovered by William Herschel at the end of the eighteenth century, did not fit into Newton's blueprint. Its observed orbit could not be accurately accounted for. Working independently of each other, Urbain-Jean-Joseph Leverrier, then an astronomer at the Office of Longitudes, and John Couch Adams, a young English astronomer, assumed the existence of an unknown planet whose mass was altering Uranus's trajectory. Reaching this conclusion had required a highly sophisticated calculation of perturbations. It was Leverrier, an astronomer at the Observatoire de Paris and its future director, who communicated the coordinates of the hypothetical planet to the astronomer J. G. Galle in Berlin. Sure enough, Galle found Neptune at almost the precise location indicated by Leverrier. Adams, for his part, had not managed to convince G. B. Airy, the director of the Greenwich Observatory, of the correctness of his calculations. The observatory at Cambridge joined the search, but they realized too late that they possessed three sightings of the planet among those of three thousand stars.[2] Given the circumstances surrounding it, the discovery of the new planet created an enormous sensation, with the press playing on the French-English rivalry. The existence of a new planet had been predicted based on a purely theoretical analysis. It was certainly one of the greatest achievements in the history of Newton's theory of gravitation. His theory was not only precise; it was true, and it had now become untouchable, a myth.

Small and situated near the sun, Mercury is a planet difficult to see. Observations of it had long lacked precision, to the point that the failure to accurately determine its orbit was not a source of concern. Due to the significant perturbations caused by the other planets, its motion is not an ellipse but an extremely complex curve. At the Observatoire de Paris, Leverrier early on tackled the problem of Mercury's motion and formulated a theory in 1843 that he subsequently refined. In 1859, he announced the discovery of an anomaly in Mercury's motion, an advance of some 38 degrees of arc per century.[3] This was in disagreement with the Newtonian prediction of the planet's motion. A minor discrepancy, to be sure, but hardly a tolerable one given the perfection of the theory's predictions. The problem could have been solved by increasing the mass of Venus—Mercury's neighbor—by 10 percent, but this increase entailed a new anomaly on Venus's other neighbor, the Earth, which was equally unacceptable. Unable to account for the motion of Mercury's perihelion without modifying the distribution of mass in the solar system,

Figure 7.1. Le Verrier pointing at Neptune. A sketch by Dupain for a ceiling at the Paris Observatory. (Photo: Observatoire de Paris)

Leverrier turned to the strategy that had led him to the discovery of Neptune.

The missing matter should affect Mercury but not Venus, whose theory of orbital motion was altogether satisfactory, so it was between Mercury and the Sun that the search for another planet—or for a procession of smaller planets such as the one just discovered between Mars and Jupiter—should be conducted. It was at that moment that the advance of 38″ of Mercury's perihelion entered the history of astronomy and later that of gravitation. Leverrier did not really believe in the existence of a single planet that, even if close to the Sun, could have escaped observation by the astronomers' sharp eyes. However, shortly after the publication of his results, he received a letter from a veterinarian and amateur astronomer who claimed to have seen the transit of a small planet between Mercury and the Sun. A cautious Leverrier then paid a visit to his correspondent, who lived southwest of Paris, in order to question him and test his astronomical equipment. He was satisfied with his investigation and the new planet was called Vulcan. Its history occupied many minds and countless telescopes. Vulcan's discovery was compared to Neptune's, and no efforts were spared to spot it, in particular during solar eclipses, but with no success.[4] It was soon realized that Vulcan alone would not suffice and that it would take sixteen planets like it to really explain the anomaly—sixteen planets as invisible as Vulcan.

In 1882, the existence of Vulcan was rejected, and Leverrier's initial conjecture, that of a ring of asteroids between Mercury and the Sun, was believed worthy of further consideration. It was then that Simon Newcomb had a fresh go at the problem by combining the observations made since the end of the seventeenth century. He showed that the unexplained advance of Mercury was in fact of 43″ per century,[5] slightly larger than Leverrier had thought. He also detected some weak anomalies of Venus's knot, the perihelion of Mars, and the eccentricity of Mercury. Among all the possible hypotheses, he favored a modification of Newton's law. Using an ad hoc force law in $1/r^n$ and choosing $n = 2.0000001574$, he managed to account for the advance of Mercury's perihelion and for the anomalies of the other planets. In 1903, it turned out that Newcomb's hypothesis would have led to serious problems regarding the Moon's perigee, and many astronomers abandoned it.

Hugo von Seeliger, an influent German astronomer and director of the Munich Observatory, started from a very ancient observation, that of the zodiacal light, which is a faint glimmer in the night sky due to the diffusion of solar light on a cloud of interplanetary dust that surrounds the Sun. The density of this zodiacal matter was not known, and Seeliger showed that its mass was enough to explain the advance of Mercury's perihelion. From 1906 on, Seeliger's hypothesis became the most commonly accepted one. It was, however, not the only one, for at the end of the nineteenth century some scientists had proposed a modification of Newton's law by a force law that depended on speed, from which they had derived a formula for the advance of the perihelion identical to that later obtained in general relativity. A number of other theories were developed to solve the problem, sometimes at a fairly high price: the introduction of totally unexciting ad hoc hypotheses.

In 1906 and 1907, Henri Poincaré gave a course at the Sorbonne, "The limits of Newton's law."[6] The class notes show that Poincaré had insisted on one of the oldest criticisms addressed to Newton's theory, the lack of a "satisfactory explanation of attraction."[7] Indeed, while Newton's theory provided an (almost) perfect description of gravitational phenomena, it did not offer any reason for the attraction.

But the conclusion of the course, typical of the time and of the difficulties Einstein would face to promote his ideas, is that Newton should prevail: "On the other hand, we have no valid reason to modify Newton's law. The most serious discrepancy is the advance of Mercury's perihelion. But this disagreement is likely due to the existence of a ring. . . ."[8]

On the question of Mercury's anomaly, Poincaré's position was ultimately no different from the one prevailing among the astronomers of the time: they hoped to solve all their problems thanks to the motion of the inferior planets.[9] "Hoped" is the right word, for not only they had to avoid digging one hole to fill up another, but they had to account for the rest of the anomalies, that of the knot of Venus, for instance, which would not be done until 1929—and by purely Newtonian considerations.

EINSTEIN AND MERCURY'S PERIHELION

The solution to the perihelion question would be a boost for the budding theory, but only a strategic one and not the magnificent solution of a crucial question. And yet, it was thanks to that first test that the theory would endure through the years 1915 to 1919, beside providing Einstein in November 1915, as we have seen, with an indication that he was on the right track. In the infinitely complex search for a new theory, everything matters, everything may be useful. First of all, it matters in relation to the scientific world, which needs to be convinced that the creator is on the right track, but it also matters to the inventor himself and in particular during periods of doubt. To be sure, the resolution of Mercury's anomaly represented, among many other elements, a sign that a solution might be at hand. But it was not impossible that things could turn out to be more complex, as was the case of Venus knot.

On Christmas Eve 1907, as he worked on the broad outline of his relativistic theory of gravitation, Einstein hoped to be able to explain the secular variations of Mercury's perihelion. Solving this anomaly was already part of his strategy but at that moment he was not optimistic. He carried on with his work and although his calculations do not survive, we can nevertheless try to understand his reasoning. Einstein knew perfectly well from his 1905 theory—which was not yet called special relativity!—that he would obtain a modified version of Newton's law, a variant of which he did not know a great deal yet except that it would only be relevant if the motions of the bodies involved were sufficiently fast, for the effect of the new kinematics only appeared at high speeds. Now, the closer a planet is to the Sun, the higher its speed, so that a relativistic effect would most likely manifest itself in the motion of Mercury, the planet nearest to the Sun. And Mercury was precisely the planet causing the greatest problem for the specialists in Newton's theory. It is highly likely that Einstein then thought along those lines and entertained hopes of explaining the advance of Mercury's perihelion in the context of a relativistic theory of gravitation. That is the reason why he started working on the problem so early on. He was still far from his goal, but he was on the right track!

In May and June of 1913, Einstein and Grossmann had put together the first version of what would become general relativity—"Outline of a Generalized Theory of Relativity and of a Theory of Gravitation."[10] With another friend, Michele Besso, Einstein then tried his new theory on the question of Mercury's anomaly. Their calculation predicted an advance of only 18 seconds of arc per century (instead of the 38″ then expected) and would not be published.[11] It was a disappointing result and one of the reasons for his giving up that preliminary version of his theory of gravitation. It is not surprising for him to have performed such a calculation, for it was the only result connecting his theoretical essays with actual observations.

Even if several relativistic theories of gravitation in embryonic form had then been proposed—by Poincaré, among others—Einstein seems to be the only one to actually have taken the trouble to calculate the advance of Mercury's perihelion. In fact, the question of Mercury's anomaly was considered by many specialists to be a "variable geometry" test, for it was always possible to assume the existence of more or less matter between Mercury and the Sun (whether zodiacal matter or small planets) to suit one's purposes.

Einstein put the perihelion problem aside until November 1915. Then, sensing that the time was ripe and fearing perhaps that someone else (Hilbert!) might be about to find the solution, he did not hesitate to publish his result (on 18 November and therefore a week before his theory was complete), emphasizing the particularly remarkable agreement between the theoretical prediction (43 seconds of arc per century) and the observed value (45″ plus or minus 5″). This quite extraordinary result gave him "great satisfaction," as he wrote to several people and in particular to Sommerfeld, to whom he confessed his regret for having had such a low opinion of astronomy: "The result of the perihelion motion of Mercury gives me great satisfaction. How helpful to us here is astronomy's pedantic accuracy, which I often used to ridicule secretly!"[12]

There is no doubt that despite the faintness of the effect, this first test gave general relativity an empirical foothold of undeniable value, especially since the theory was still in its infancy. However, this result would take on its full significance only after

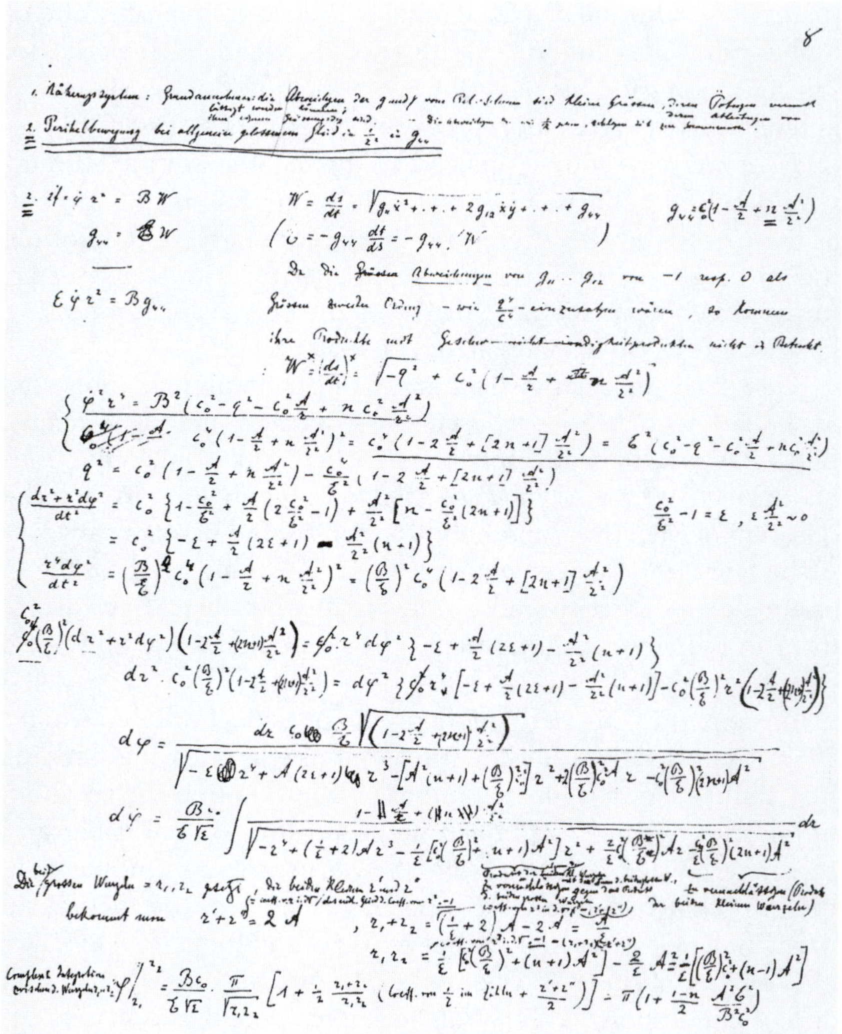

Figure 7.2. Calculations of the motion of Mercury's perihelion. (From Einstein and Besso, 1913, *Collected Papers of Albert Einstein* (CPAE), vol. 4, p. 644)

the results of the 1919 eclipse, which would dismiss Newton's theory for good. Both the interest of the first test and the crucial importance of the second should be noted. The first helped to bolster the second, which would be the main support of the theory until 1960,[13] when the third test, the line shifts, would at last be clearly measured by Pound and Rebka.

Einstein now had an *explanation* for the advance of Mercury's perihelion or, more precisely, the advance followed from his equations. We will now try to understand how his theory works: not so much how Einstein carried out his calculations, but rather how a theory like his can predict such an effect. We begin by focusing on the field equations as they are understood today.

THE RELATIVIST AT WORK

To talk about a physical theory so pervaded by mathematics as general relativity is a very delicate task, for as soon as we depart from the equations we inevitably risk talking nonsense. And yet, too much caution in our speech and an absence of images would render the whole thing unintelligible. For equations alone are not enough: they need interpreters who speak in their mother tongue before speaking the language of equations and also speak "with their hands," the specialty of physicists.

Far from being a geometrization of physics, general relativity brought about a *physicalization* (if you would allow me this neologism) of geometry, a physical chrono-geometry. Axiomatic geometry became (again) physical geometry (see chapter 6). It is not an exaggeration to say that general relativity is the current theory of the structure of space, the theory of *physical* geometry. Geometry has become a physical science: the structure of space-time depended on the matter it contains. This is precisely what the field equations, $E_{\mu\nu} = \chi T_{\mu\nu}$ that Einstein achieved at the end of 1915, express: while E describes the geometry, T represents the material components of the intended model, and it is from right to left that we must read and understand this equation: $T \rightarrow E$, the matter present determines the chrono-geometric or physical-geometric structure of space-time. From the matter present, which is given a priori, the geometry of space-time is derived.

We must emphasize here the leap, the epistemological break that underlies this truly radical vision. For until then, that is, in the Newtonian theory of gravitation, space—absolute space—was given beforehand, once and for all, forever. An attribute of God, space was certainly not contingent, and it could not therefore be an *object* of physics.

The technical practice of general relativity has greatly developed since 1915. For the benefit of those readers who never had the opportunity to manipulate the theory, I will summarize here, as simply as possible, the stages in the history of this practice as they are currently perceived.

Before examining the work of a relativist who sets out to determine the structure of a given gravitational system, it is perhaps useful to recall what a mechanist, working in the context of Newton's theory, must achieve if he wants to determine all possible trajectories around the Sun.

Our Newtonian mechanist has at his disposal an absolute space at whose center he has placed a spherical mass, while a no less absolute time goes by. He is going to apply to his test particle, the Earth, say, the fundamental law of dynamics. The Earth is subject to an action: a force of gravity proportional to the product of the (gravitational) masses of the Sun and the Earth according to Newton's law of gravitation: $F = -GMm/r^2$. On the other hand, the Earth opposes (through its inertial mass, m_0) its inertia to motion and it is subject to an acceleration a: $F = m_0\, a$. Equating these forces, our mechanist obtains a very simple equation which, thanks to the equivalence between the Earth's gravitational and inertial mass ($m = m_0$), is written:

$$-GM/r^2 = a.$$

This equation is really quite easy to integrate, and from it we can derive in particular Kepler's law giving the shape of the trajectory: an ellipse, with the Sun at one of its foci.

Let us follow step by step the relativist who wishes to determine the structure of a given gravitational system and the trajectories of the test bodies in motion. What is the structure of space-time around a spherical or point-like mass? What is in this field—the Sun's—the planetary orbits and the trajectories of light rays? To answer these questions, he has available a Riemann manifold described by the space-time line element, proper time:

$$ds^2 = g_{\mu\nu} dx^\mu dx^\nu$$

where μ and ν take on the values 1, 2, 3, and 4. He also has Einstein's field equations written above in a tensorial form so condensed that it conceals their complexity. Once developed, these equations become a horrendously complicated system of second-order partial differential equations where the ten $g_{\mu\nu}$ (10 and not $16 = 4 \times 4$, as one would expect, for the $g_{\mu\nu}$ are symmetric: $g_{\mu\nu} = g_{\nu\mu}$) are the unknowns. These equations involve the first- and second-order partial derivatives of the $g_{\mu\nu}$.

These equations specify the structure of the manifold because they determine the form of the $g_{\mu\nu}(x_\alpha)$, called gravitational potentials, as a function of the coordinates $x_1, \ldots, x_4$ of space-time. The left-hand member $E_{\mu\nu}$ is a nonlinear function of the $g_{\mu\nu}$ and of its first derivatives $\partial_\alpha g_{\mu\nu}$, and a linear function of its second derivatives $\partial_\alpha \partial_\beta g_{\mu\nu}$. We must be given a priori the form of the right-hand member, the matter tensor $T_{\mu\nu}$ that describes locally the sources of the gravitational field: χ is simply the universal gravitational constant.

In the present example, the gravitational field created by the Sun, this matter tensor $T_{\mu\nu}$ is zero because we are only looking for the structure of (empty) space around a mass M that we may suppose to be either a point or spherical (from a spatial point of view). The external solution will be the same in both cases.

We now have to solve these field equations, either exactly or approximately. This generally involves solving ten differential equations for ten unknown gravitational potentials, which, thanks to the freedom (covariance!) afforded by the free choice of the four coordinates, are reduced to six. In order to state and

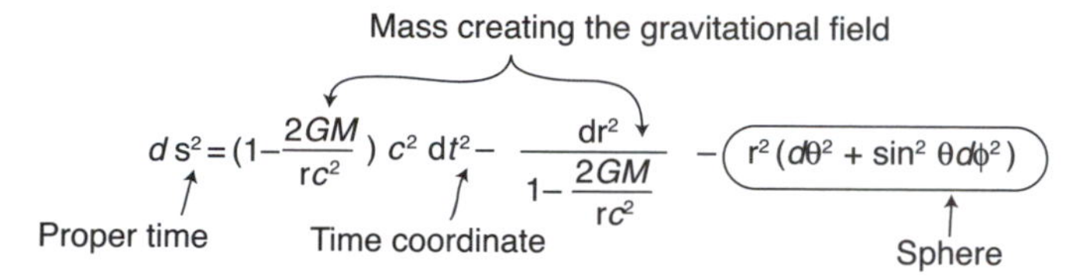

Figure 7.3. The metric of Schwarzschild's solution in its "classical" form is in fact due to J. Droste, 1916.

later solve these field equations, we must first choose the coordinates in a suitable way, depending on the symmetries of the problem at hand. The invariance of the problem with respect to a spherical symmetry *naturally* leads us to select polar coordinates r, θ, ψ, the fourth coordinate denoted, no less naturally, by t. There are, finally, only two unknown functions $g_{11}(r, t)$ and $g_{44}(r, t)$ that are solutions of two second-order differential equations. The solution, known as Schwarzschild's solution (after its author) is then easily obtained.

Equipped with the (exact) solution of the field equations, that is, knowing the form of the $g_{\mu\nu}$, we will now be able to study the shape of space-time, which is bent, structured, by our physical hypotheses (matter present and symmetry).

This is not the method employed by Einstein in his November 1915 article. He did not believe that the exact solution was so simple and preferred to assume that space was almost Euclidean, almost *flat*, and therefore that the gravitational potentials were very similar to those in special relativity (i.e., close to 1 or –1). In fact, Einstein assumed the so-called weak field hypothesis (which in this case becomes $GM/rc^2 << 1$) and worked with an approximate metric in which the potentials were replaced by their post-Newtonian expansion in GM/rc^2, of the form $1 + \alpha GM/rc^2$. He carried out what is now known as "a post-Newtonian analysis."

As for us, we can now study any given astronomical problem using the solution just found. This requires determining and solving the equations of the trajectory of test bodies which, according to the principles of the theory, are represented by the geodesics. In the case of a light particle, these geodesics are called *isotropic* (they have then zero length), and *time* geodesics, if the particle is a material one. Thanks to a suitably chosen approximation, such a study enables us to determine the orbits and to predict the deviations of the new theory with respect to the old one. In the case of Schwarzschild's solution, we shall easily find the relativistic advance of the perihelion of a planet, the advance *relative* to the Newtonian theory.

These three stages are sufficient to obtain general relativity's post-Newtonian predictions. And, in the above example, that is all that is needed to determine the theoretical predictions for two of general relativity's three classical tests: the advance of Mercury's perihelion and the deflection of light rays near the Sun. That is, in broad outline, the way in which Schwarzschild's space-time has been interpreted and taught during fifty years. But there is much more to be learned about the structure of space.

To help the reader understand, we shall consider two questions in which trajectories play a major role: that of the physical observables in general relativity and that of the shape and the limits of space.

Everything would be relatively simple had the theory provided from the start a set of observable physical magnitudes defined by invariants—that is, intrinsically defined—that could be expressed in any chosen coordinate system. But such was not the case. In 1916, general relativity did not provide any answer to these questions, and no intrinsic definition of an observable magnitude, except that of proper time, s, was then known. Of course, the notions of distance and force have been abandoned, but (and this is seldom explicitly mentioned in the textbooks) we no longer know what in general the speeds of two distant particles are or their relative acceleration; we do not know what mass or energy are either. So then, how are we supposed to make the connection between the observed physical quantities—the *observables*—and the theory?

Actually, and curiously enough, we do not need much, at least regarding the question of the advance of the perihelion. Once we have a model of the space-time representing the solar system, *it is enough to ask ourselves what it is we are really observing*.

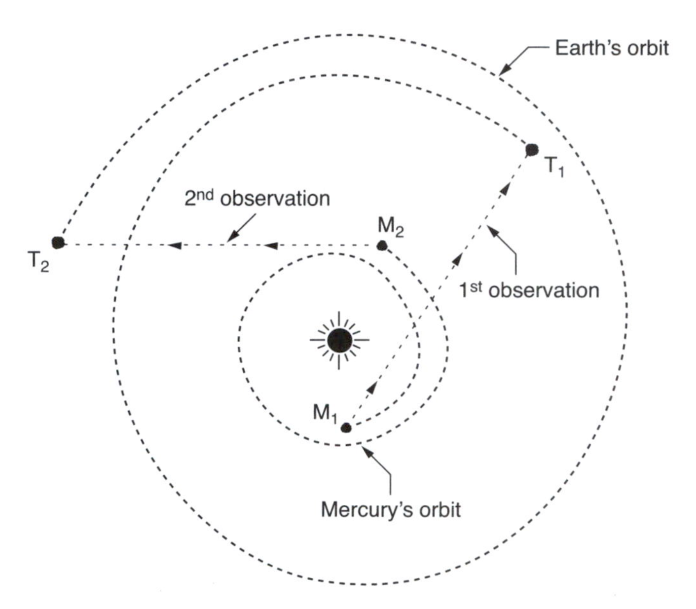

Figure 7.4. Diagram illustrating the elements of an orbit.

What do we mean by an observation of Mercury? It is simply an astronomical sight that lines up the observer, Mercury, and a fixed star at a very precise moment. It is again light that makes the measurement possible, that is, the trajectory of a light ray joining Mercury (M_1) and our observer (T_1). Assuming that the two measurements used to calculate the variation in the position of Mercury are taken with the same instrument and at the same place, the observer has described, along with the Earth, another trajectory whose exact equation can be obtained, and whose boundary conditions coincide with the arrival of the photons involved in both measurements.

In short, all we need is a model of space-time representing the solar system, trajectories, and co-occurrence: events consisting of intersections between trajectories. All in all, we must put together end-to-end four trajectories that follow on each other: the trajectory M_1T_1 of the light particle that reached our observatory during the first sight; its orbit, that is, the trajectory T_1T_2 of the Earth between the first and the second observation; Mercury's orbit M_1M_2 between the two observations; and, to close up the loop, the photon traveling from Mercury to our observatory during the second sight, M_2T_2.

The proof of Mercury's passage through its perihelion (or some other point) is provided by the photons reflected by Mercury at that instant (or by some other co-occurrence, for example its passage in front of a fixed star). Since we cannot neglect the motion of the observer on Earth, we need four equations of motion, four trajectories: Mercury's, the observer's between the initial and final instants, and the two equations of the light rays carrying the information of Mercury's position to Earth. That is how we can calculate the time between the two terrestrial observations corresponding to the two consecutive passages of the planet through its perihelion, and then compare it with the observed time and the Newtonian calculations.

NOTES

1. On the history of Mercury's perihelion, cf. Roseveare, 1982.

2. J. Lévy, in Taton, 1961, volume 3, p. 150.

3. In order to study the planetary motions and draw up a table of their observations, it is necessary to choose a fixed point in their orbit. This may be either the planet's perihelion—the point in its orbit closest to the Sun—or one of the planet's knots, which is the intersection of the plane containing the planet's orbit with the ecliptic (the plane in which the Earth revolves around the Sun). It is then possible to compare the observations with the theoretical calculations and so attempt to "account for" the real motion. If the observations do not agree with the Newtonian picture, we have an "anomaly." This may be either an unexplained motion of the perihelion or one of the knots (advance or retardment), or a disagreement on the value of one of the orbit's parameters, the mass, for instance, or the eccentricity of the ellipse.

Notice the extraordinary accuracy of Newton's theory in accounting for the planetary motions: 38 seconds of arc per century! And yet, Mercury's position is not exactly that predicted by the theory. This "not exactly" is rather slim: if, after one century, we look for Mercury at the place the theory tells us to look, we shall find it, but the planet will be 38 seconds (actually, 43, as we shall see) off the theoretical position.

This is less than one minute of arc, that is, an angle of one sixtieth of a degree. A minimal difference which Newtonian astronomers cannot put up with, such is their conviction that their theory is right. And, as we shall see, it is this tiny angle that general relativity will first account for—peanuts! In this regard, consider Einstein's remark to Sommerfeld (later in the chapter) on astronomy's "pedantic accuracy."

4. Vulcan, an inferior planet, would have to be near the Sun, and therefore it would have been difficult to see except during solar eclipses, the only occasions on which we can observe what goes on near the Sun.

5. Based on the discussion and the results regarding Mercury's transit from 1667 to 1881. Newcomb, 1882, p. 473. In this respect, cf. also Roseveare, 1982, p. 41.

6. Poincaré, 1953.

7. Ibid., 1953, p. 265. Quoted by Katzir, 1996, p. 12.

8. Ibid.

9. These are the planets that are closer to the Sun than the Earth.

10. Einstein, 1913, *CPAE*, vol. 4, pp. 302–343.

11. It was of course published in the collected papers, *CPAE*, vol. 4, pp. 360–473.

12. Einstein to A. Sommerfeld, 9 December 1915, *CPAE*, vol. 8, p. 217; English translation, vol. 8, p. 159.

13. I shall skip here the history of the difficult verification of this test. In this connection, see chapter 10.

CHAPTER EIGHT

Relativity *Verified*: The Deflection of Light Rays

In 1907, Einstein had predicted that in a theory that would properly combine special relativity and gravitation, the trajectory of light would be curved (by gravitation). In fact, his idea was not completely new. Already in 1801, the Austrian astronomer Johann von Soldner had studied the question of the effect of gravitation on light and had calculated (based on Newton's theory) the deflection of a light ray "which passes close by a celestial body."[1] Soldner's contribution was not unique. John Michell, an English physicist, had also studied the expected influence of gravitation on light.[2] In Michell's view, Newton's theory should be truly universal. Didn't it apply to all material bodies, not only throughout the solar system but also in the twin stars whose existence he had predicted in 1767? Why should a light particle escape this force, this mechanics? He then showed the possible existence of *dark bodies*, as Laplace would later call them, which are in a way the ancestors of black holes.

Newton's corpuscular optics dominated the eighteenth century, and it was thought that light was composed of small material particles traveling very fast. His corpuscular theory of refraction was based on a sort of ballistics of light. If light particles were refracted by glass, it was because of the presence, on the crystal's surface, of an extremely powerful force field, a *refringent atmosphere*. This was in fact a dynamics of light similar to that of gravitation. Only the nature of the force was different, and for a while the possibility that gravitation itself was at work, acting at short range, was even envisaged. The light particle refracted by that refringent atmosphere was compared to a musket ball shot horizontally whose trajectory would be gradually bent by the Earth's gravitational field.

According to Newtonian corpuscular optics, the light particle was accelerated by the refringent force, which implied that light traveled faster in glass than it did in air. Descartes' refraction law, or the law of sines, was thus recovered, and the coefficient of refraction of a given crystal, which is proportional to the speeds (of incidence and refraction) of light, could be calculated. All this worked reasonably well until the beginning of the nineteenth century.

Michell worked hard to come up with a purely Newtonian theory of the effect of gravitation on light particles. Insofar as this effect was independent of the mass of the body subject to the field (the principle of equivalence!), why wouldn't it be exerted on light as well? It was known that light traveled very fast, but it was not possible to assume that its speed was constant, even if all observations pointed in that direction, in particular those concerning aberration. The speed of light could not be constant simply because in Newtonian mechanics a light particle, just as any other particle, would have to obey the law of addition of velocities (see chapter 1).

The argument proposed by Soldner was then highly plausible. If, according to Michell, light were subject to gravitation just as any other material particle, then the trajectory of a light particle traveling near the Sun should curve. Soldner's calculations were perfect, just as Michell's had been twenty years earlier. They explicitly predicted the maximum deflection of a light ray grazing the Sun to be 0.84 second of arc, and that due to the Earth's gravitational field to be 0.001 second of arc. However, all that was far too small to be observed or to have any effect on astronomical observations.

After the corpuscular theory of light was abandoned in favor of the wave theory at the turn of the nineteenth century, such an effect was no longer possible, for it was unconceivable that gravitation could affect a light wave. And beside, there was no reason to believe otherwise. On the contrary, all observations suggested that neither the trajectory of light, always a straight line, nor its speed, always constant, were affected by gravitation.

Figure 8.1. Johann von Soldner (1776–1833), Prussian astronomer. In 1801 he calculated the deflection of light rays due to the Sun's gravitational field. (Deutsches Museum München)

MEASURING THE DEFLECTION

The idea that gravitation could affect light was simply forgotten until Einstein revived it in 1907 (as we have seen in chapter 5) and again in the spring of 1911. Wasn't he working on a theory

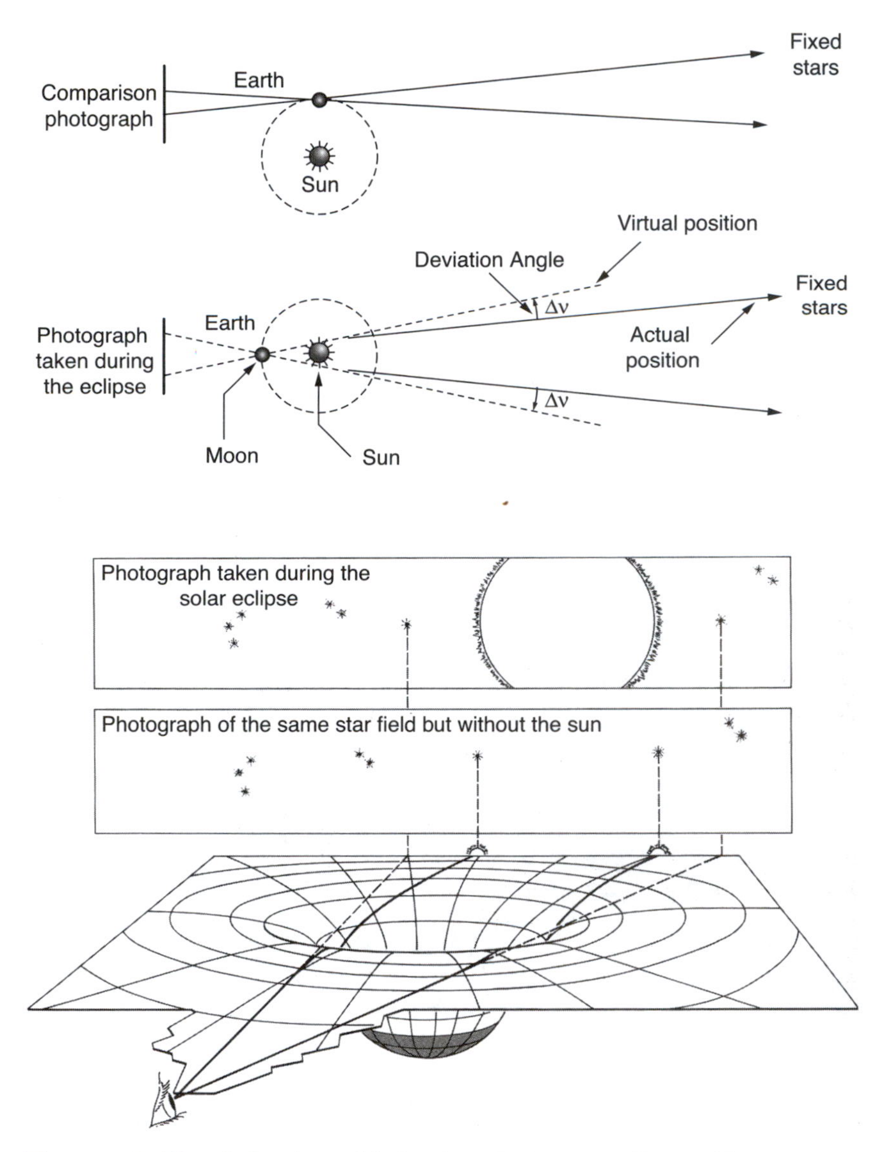

Figure 8.2. The deflection of light rays during an eclipse. *Above*: a classical diagram. *Below*: a diagram in curved space-time. (From Thorne, Misner, Wheeler, 1973)

of gravitation where light played a crucial role? But he was unaware of Michell's and Soldner's results; he only knew that it was a natural idea in Newtonian optics.[3]

Einstein's hypotheses and calculations of 1911 were not essentially different from Soldner's, but the goals of the two men could not have been farther apart. Surely, the rediscovery in 1921 of Soldner's precursory calculations would, in the anti-Semitic atmosphere of the time, inevitably raise some absurd doubts and provoke a regrettable debate led by Philipp Lenard. After calculating the magnitude of the effect, Einstein explained in his article how it would be possible to observe it.[4] A light ray coming from a star and passing near the Sun, a planet, or other massive body will be slightly bent, deflected. And the apparent position of the star will be different from what it would have been in the absence of this massive body. Therefore, the appearance of a star field should change as it travels near the Sun or a planet. It is like looking through an old window glass: the light rays, deflected by the flaws and the bubbles in the glass, will distort the image of the postman delivering the mail.

But how can such an effect be measured? The Sun, with its relatively strong gravitational field would be an ideal candidate, but its glare will mask the effect we are trying to measure: the star field near its edge. For the deflection is the greatest for a light ray grazing the Sun (see figure 8.2), but it rapidly decreases if the ray is further away from its center. However, it is perhaps possible to observe the effect during an eclipse, when the hidden Sun allows us to catch sight of those stars closest to its edge. The shift in position of the image of the most deflected stars is minuscule, of the same magnitude as the definition of the photograph. Measuring positions directly from the plate is therefore out of the question, because they are not accurate enough. But it is not impossible to compare a photograph taken during an eclipse to another one taken later, in the absence of the Sun. The possible variation in the position of the stars form one photograph to the other could enable us to measure the effect. The deflection of light by a gravitational field could then be demonstrated and its magnitude measured. Einstein wondered whether the effect would be visible during the day near Jupiter, the largest planet in the solar system. But the effect, one hundred times weaker than the one involving the Sun, would be undetectable.

At the end of his article, Einstein expressed his wish that astronomers could carry out these observations, and even before the publication of his article in 1911, in *Annalen der Physik,* he had an opportunity to advance his cause. Erwin Freundlich, an assistant at the Royal Observatory of Prussia in Berlin, got wind of the article—and of Einstein's interest in astronomy—thanks to the visit of one of Einstein's colleagues, a certain Leo Wentzel Pollak, an instructor at the German University of Prague, where Einstein himself was teaching. Convinced and excited, Freundlich wrote to Einstein the very same day of this visit.

Einstein replied without delay: "If only we had a truly larger planet than Jupiter! But Nature did not deem it her business to make the discovery of her laws easy for us."[5] And he added: "But one thing, nevertheless, can be stated with certainty: If such a deflection does not exist, the assumptions of the theory are not correct."[6]

Freundlich believed he could solve the problem fairly easily; he was not short of ideas. He even thought it would be possible to observe in broad daylight the fixed stars near the edge of the Sun. But Einstein doubted that things could be so simple and asked the opinion of the director of the Mount Wilson solar observatory in California, who strongly recommended performing the measurements during an eclipse, for that would eliminate plenty of difficulties.

Freundlich redoubled his efforts to convince his colleagues of the importance of those measurements. In October 1911 he met with C. D. Perrine, the former astronomer at the Lick Observatory and the current director of the Observatory of Cordoba, in Argentina, who was passing through Berlin. To measure the deflection, Freundlich expected to be able to use the photographic plates from previous eclipses. William Wallace Campbell, director of the Lick Observatory, sent him copies of photographs taken during the most recent eclipses as well as photographs to compare them with. Unfortunately, those plates, taken for the purpose of studying the solar corona or to search for planets near the Sun (remember that, for the same reasons, the observation of Vulcan required an eclipse) were not sufficiently sharp to permit the measurements: first, because during those eclipses the lens had to follow the Sun rather than the fixed stars, which then traced out a small arc whose position Freundlich could not precisely measure; second, because the Sun was not at the center of the photographs, this made the measurements all the more difficult to perform. The next move was then the preparation of an astronomical expedition to the site of an eclipse to try to measure the deflection.

THE FIRST ATTEMPT

On his return to Argentina, Perrine added measurements of the deflection of light to the program of observations for the next eclipse. The shadow of the Moon would sweep the South American continent on 10 October 1912, and the eclipse would be total on a narrow strip across Brazil. Perrine set up camp at Christina, in the Minas Gerais, equipped with the material lent to him by Campbell. Other than the deflection of light rays, there were many reasons to observe a total solar eclipse, notably the study of the solar corona, so no less than eight expeditions were present along the totality line. At Passa Quatro was the Brazilian mission headed by Henrique Morize, director of the Rio de Janeiro Observatory; the French delegation under the direction of M. Stephanik; as well as the British mission, which included Charles Davidson, from the Greenwich Observatory and an experienced observer of these phenomena, and Arthur Eddington, from the same observatory. Alas, the rain frustrated everybody's efforts, in particular Perrine's, who was the only one trying to measure the deflection. He was already thinking of an exceptional eclipse that would cross Brazil in seven years time, and he advised Morize to search for the best possible observation site.

Clouds and rain are of course an astronomer's worst enemies at all times, but even more so during an eclipse, because of the considerable sums invested for a few minutes of observation. This is the reason why the statistics and the weather conditions must be carefully estimated before deciding on the sighting location. The weather forecast is the most important factor, but the quality of the photographs will also depend on the magnitude of the atmospheric refraction and hence on the height of the Sun above the horizon, which varies with the observation site.[7] Finally, the accessibility of the place is another factor when estimating the overall cost of the mission. As far as the measurement of the deflection of light near the Sun is concerned, the richness of the star field on which the eclipsed Sun will be projected is vital. If this projection includes a constellation rich in very bright stars that are relatively well spread, the pictures will obviously be more interesting than those of a section of the sky containing only a few, dim stars lumped together, for it is the apparent shift in position of those stars from one photograph to another that is compared.

As for Freundlich, he was busy trying to raise funds for the observation of the next total eclipse, in Russia on 21 August 1914. The prospects were so dim after the refusal of Hermann Struve, director of the Berlin Observatory, that for a while Einstein contemplated making a personal contribution of two thousand marks. This was not necessary after all, for through Einstein's intervention private funds were secured, in particular from the Krupp family. Freundlich was not alone in Crimea, the Lick Observatory having sent an important delegation headed by its director, Campbell, and including Heber D. Curtis, Campbell's assistant, who was specially charged with measuring the deflection. The equipment (the objectives are those used for the search of planets near the Sun) was from the Lick Observatory and appeared well suited to the task. But once again, it rained in Brovary, where they were located. Apparently, the observation site had been poorly chosen. In his report on the expedition, Campbell was particularly critical in this respect. It is a juicy document, although of little scientific interest given that no observation took place. But the expenses incurred had to be justified, because the funds came from a private donation, and the work of the astronomers had to be justified, too, through assurances that everything (or almost) had been done to ensure that the expedition would be as successful as possible. Campbell bitterly remarked that Davidson had no problem with the weather and was able to observe the eclipse in Minsk, but the deflection of light was not on the program of that English expedition.

Perrine was also in Crimea, where he met up with Freundlich. But on 28 July 1914, one month after the assassination of the Archduke Franz Ferdinand, Austro-Hungary invaded Serbia. The mobilization order was soon given in Russia, and Freundlich and his colleagues, who were in Theodosia, were not able to set up their gear. As members of the German army reserve, they were taken prisoner, for Germany had just declared war on Russia. In Brovary, the Americans had no problem with the authorities: Campbell had asked the English consul for "two policemen in uniform . . . to guard the expedition against the acts of ignorant or excited people who may be inclined to connect the eclipse with events occurring around them."[8] Orders read by the *ispravnik*, an official in charge of the district, warned the local population that children must remain indoors and no cattle should be left out in the fields, the authorities fearing that the peasants might imagine a connection between the preparations for war, the American presence, and the eclipse. The return to Lick was not easy, and the instruments were stored at the Poulkowo National Observatory for a long time.

All this did not help Curtis with the preparations for the next eclipse, which could be observed not far from Lick, at Goldengale, in Washington State, on 8 June 1918. Because of the war, the instruments stored in Crimea were not be back in time, and Curtis had to make do with objectives lent by the Oakland Observatory, which were less suitable to this particular mission. The weather was not ideal at the time of the eclipse but, between two clouds, a few photographs were taken, the first ones specially shot to detect Einstein's deflection. The measurements on these plates would only begin in July 1919, for in the meantime Curtis had been drafted. Moreover, there were other problems with the photographs. The images were not clear enough, probably due to a deficient installation that made the telescope unstable during the shots. All this was discussed, bitterly and at length, by Curtis and

Campbell. Measurements were made and repeated several times. The results were not conclusive, but Curtis interpreted them as showing no effect, thus confirming what he had expected all along. Certain American astronomers were not unhappy with this outcome, in particular George Ellery Hale, director of Mount Wilson, who wrote to Curtis sending his "hearty congratulations on the result of your eclipse work," and who confessed to be "much pleased to hear that you find no evidence of the existence of the Einstein effect."[9] In fact, it is the theory itself, especially its mathematical apparatus, that those astronomers could not accept, as Curtis himself admitted in a short 1917 text, in which he candidly observed that "the mathematics of such a physical universe is somewhat complicated, but it seems to fit well with all observed phenomena."[10] He passed quickly over this aspect of Einstein's theory before expressing, not without some irony, an uneasiness clearly shared by many of his colleagues: "Many will feel that the idea of a four-dimensional space-time is fully as difficult of comprehension as was the mystery of gravitation, all-pervading, inexplicable, in our classical physical theories. While the mathematician is willing to admit that many other forms of space or geometries of space would satisfy physical science as well as the Euclidean, we must confess that we are still of the point of view of the mathematician who stated that, while it would be possible, in a four-dimensioned universe, to turn an egg inside out without breaking its shell, still he realized that there were many practical difficulties in the way of the accomplishment of this feat."[11]

While Campbell was in London, in July 1919, the English expeditions were about to return from the southern hemisphere with the much-awaited photographs. At a meeting of the Royal Astronomical Society, Campbell said a few words, which he would later bitterly regret, on the Goldengale results: "It is my opinion that Dr. Curtis's results preclude the larger Einstein effect, but not the smaller amount expected according to the original Einstein hypothesis."[12] As far as the Lick astronomers were concerned, general relativity had been refuted.

In fact, Campbell's analysis was based on the way Eddington had posed the question before leaving Cambridge to travel to

Africa: Is there any effect at all? Is it of the order predicted by Newton's theory, or will it confirm the predictions of Einstein's theory?

In order to properly understand the sense of all this and of the discussions that followed, we must go back to the evolution of the theoretical debate. At the end of 1915, Einstein had calculated the deflection once again, this time in the context of his new theory of general relativity. And in that famous article where he discussed the advance of Mercury's perihelion, he also gave the value of the expected deflection at the limb, which is no longer 87 seconds of arc, as his first (Newtonian) calculation suggested, but twice as much: 1′75″.

Eddington was one of the few physicists to recall Michell's theory (see chapter 12), and he therefore knew perfectly well that Newton's theory might predict such a deflection, as long as one accepted that light particles are subject to the gravitational force. If that were the case, the deflection was then the one calculated by Einstein in 1911. But Eddington was also well aware that in Newton's theory one was not obliged to assume that light is affected by gravitation, a fact that physicists had exploited since the end of the eighteenth century. Let us note in passing that it is amazing that a theory so well built, so complete and closed as Newton's, could allow such ambiguities: that light may or may not be subject to gravity. Whatever the case, in the wave approach (Fresnel's, Maxwell's, or Lorentz's) that prevailed since the beginning of the nineteenth century, the question did not arise: light and gravity could not interact.

Even if in his heart he was definitely in favor of Einstein's theory, Eddington remained pragmatic and hoped that the measurement of the deflection of light during an eclipse would settle the question. Before leaving for Principe, he prepared his analysis of the results: "There might be no deflection at all; that is to say, light might not be subject to gravitation. There might be a "half-deflection," signifying that light was subject to gravitation, as Newton has suggested, and obeyed the simple Newtonian law. Or there might be a "full deflection," confirming Einstein's instead of Newton's law."[13]

Eddington occupies a prominent place in the history of astrophysics as well as in that of Einstein's theory of gravitation. He was an extremely bright man, passionate and extroverted, who excelled at popularization; all in all, an astonishing, remarkable character. In 1905, he left Cambridge, Newton's university, to which he would return in 1913 to teach astronomy as Plumian Professor, for the Greenwich Observatory.

It is highly likely that it was in Brazil in 1912, at Passa Quatro, that Eddington first heard, from Perrine, about the theory of gravitation in progress and the deflection of light effect. In his report, the scientific results were, of course, slim, but he thanked the authorities and wrote about the practical problems of transportation, the delays, the tourist's wonder at his arrival in South America, the small local train, and the fully loaded wooden-wheeled carriage, pulled by oxen. He described the eclipse itself and the nearby forest, but above all, he expressed his bitter disappointment for an unlucky, failed expedition.

Eddington was therefore ready to receive Einstein's latest work. In 1916, he was in touch with Willem de Sitter, professor of astronomy at the Leyden Observatory in the Netherlands, who in October 1916 wrote a short article in *The Observatory*, a journal for well-informed amateur astronomers. It was a serious article, but a bit too difficult. In fact, de Sitter was inspired by Einstein's recently published paper, and he closely followed the structure and the logic of the latter. The whole artillery of general relativity was displayed on those unfortunate eight pages that the average reader (and no doubt also the advanced one) had a hard time deciphering. We cannot blame de Sitter for having written such a paper, both necessary and (too) sufficient. Someone had to try to draw attention to a theory that was hard to understand and even more difficult to explain—handicaps that would persist for a long time. We can appreciate, by contrast, how helpful Eddington would be by emphasizing the tests that might confirm the theory and by his use of images and analogies to try to get everybody, physicists as well as astronomers and even laypeople, to understand this truly subtle theory.

De Sitter's article prompted a letter from a slightly pontifical American astronomer, who complained that de Sitter "so completely passes over all *physical considerations* as actually to convey the impression that gravitation is not a *physical problem*, but only an *analytical* one."[14] He concluded with some crudely antirelativistic considerations that were then rather common: "It is the belief of many experienced investigators that the whole doctrine of relativity rests on a false basis, and will some day be cited as an illustration of foundations laid in quicksand. Dozens of books have appeared on the subject. Thus our problem to-day is not merely to discover Truth, but to discover a way out of an ensnaring mesh of errors."[15]

This vicious letter provoked an ambiguous response from James H. Jeans, an English theoretical physicist, who feared that "Einstein's theory may meet with an unfavorable reception on account of the somewhat metaphysical—one might almost say mystical—form in which his results have been expressed"; and he observed that "the more concrete part of Einstein's work is quite independent of the metaphysical garment in which it has been clad."[16] It was a defense of Einstein that nevertheless drowned the deepest meaning of his work.

THE 1919 ECLIPSE

Eddington managed to convince the Astronomer Royal, Frank Dyson, from the Greenwich Observatory, who was relatively open to observational questions but rather skeptical on the theoretical level. So, early on, Dyson called attention to the interest that the 29 May 1919 eclipse represented for the validation of Einstein's theory.

To properly understand why the 1919 eclipse represented such a unique opportunity, we must return to the analysis of the observations to be made. Two series of photographs of the star field around the Sun must be taken, the first one during the eclipse and the second before or after it but at night and under

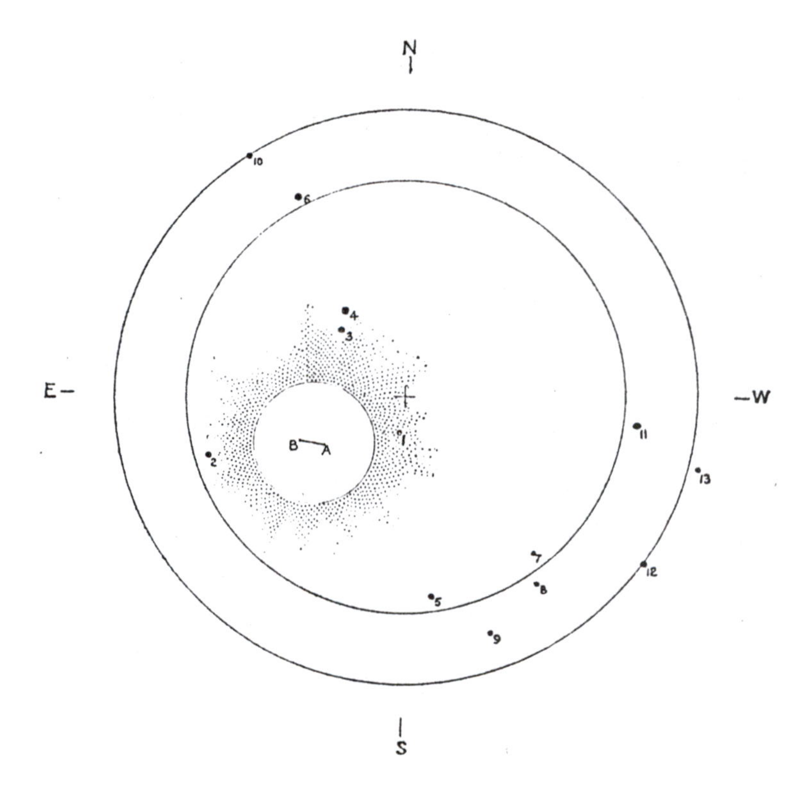

No.	Name.	Photog. Mag.	Dist. from Sun's Centre.	Displacement.
		m	′	″
1	B.D. 21° 641	7·0	23	1·20
2	Piazzi IV. 82	5·8	29	0·95
3	κ^2 Tauri	5·5	32	0·86
4	κ^1 Tauri	4·5	37	0·75
5	Piazzi IV. 61	6·0	55	0·50
6	υ Tauri	4·5	69	0·40
7	B.D. 20° 741	7·0	69	0·40
8	B.D. 20° 740	7·0	71	0·39
9	Piazzi IV. 53	7·0	71	0·39
10	72 Tauri	5·5	82	0·33
11	56 Tauri	5·5	88	0·31
12	53 Tauri	5·5	96	0·29
13	51 Tauri	6·0	105	0·26

Figure 8.3. The May 1919 eclipse field of stars. (From F. Dyson et al., 1917)

conditions as close as possible to those of the first series. In superposing the two photographs one would then be able to measure the shift (if there is a shift) in the position of each star from one picture to the other. The nearer a star is to the edge of the Sun, the limb, the more significant the shift will be. And, of course, the more stars there are and the brighter they are, the easier and the more precise the measurements. In particular, a statistical analysis of the deflection as a function of the position of each star will be necessary, and the results will be all the more precise if the stars are well scattered around the Sun. The most interesting eclipses for this kind of measurements are those in which the Sun crosses a field rich in well-scattered stars. That would be precisely the case in 1919, when the Sun would go through the Hyades, a group of stars in the head of the constellation Taurus, which contains a large number of bright stars.

At the meeting of the Joint Permanent Eclipse Committee, in November 1917, the Astronomer Royal observed that such favorable occasions were extremely rare and that there would not be another propitious eclipse for many years. To begin with, the committee proposed to send, if possible, two different expeditions. This was no ordinary decision, since they were at war, and for a country to send two simultaneous expeditions to carry out the same observations was highly exceptional. The only task of both missions was to measure Einstein's deflection. The vigor of this scientific policy decision must be credited to Dyson's authority and to Eddington's conviction regarding the significance of what was at stake.

The path of the eclipse would sweep across Africa, the Atlantic Ocean, and South America. A first expedition composed of Eddington and Cottingham would be based in Principe Island on the African coast; the other, composed of Davidson and A.C.D. Crommelin, at Sobral, a small village in northeastern Brazil. A subcommittee that included Dyson and Eddington was charged with the material organization. They were allocated a sum of eleven hundred pounds, one hundred for the equipment and one thousand for the expeditions themselves.

The choice of the instruments was of course crucial. The Cam-

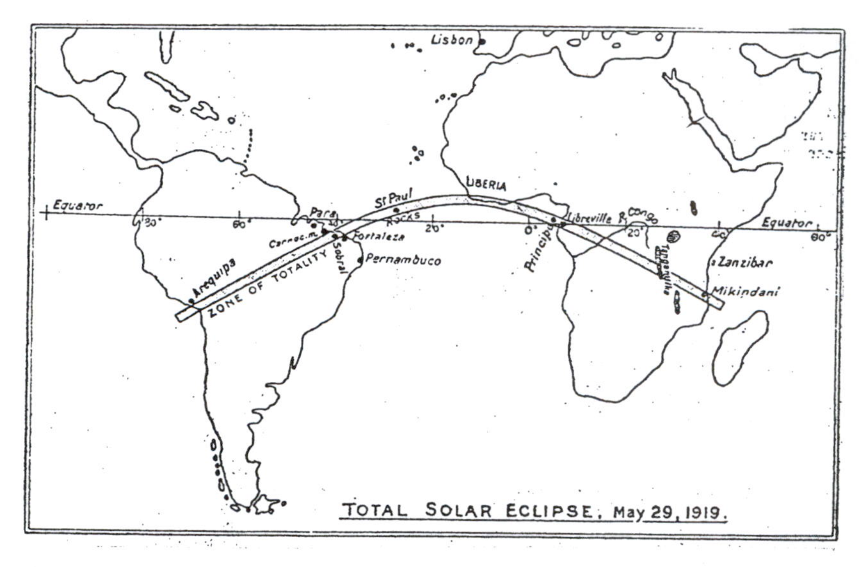

Figure 8.4. The path of the 1919 eclipse. (From A.C.D. Crommelin, 1919. Reprinted with permission from *Nature* [no. 102, 1919]. © 1919, Macmillan Magazines Ltd.)

bridge team (Eddington and Cottingham) would take with them Oxford's astrograph refracting telescope, while the other team (Davidson and Crommelin) would use that of Greenwich's at Sobral. They would also take with them a telescope with a four-inch aperture used in Sweden during a previous eclipse and the Royal Irish Society's coelostat. For these expeditions, it was preferable to choose a fixed telescope combined with a coelostat—essentially, a mirror driven by a falling weight—to reflect starlight into the telescope, where it was easily photographed.

The reasons for Eddington's participation in the expedition were not exclusively scientific. He was a devout Quaker and, as such, a conscientious objector, as everybody knew. In those difficult times, not to enlist as a volunteer to go to the front was regarded very unfavorably, and the Cambridge professors, in particular Sir Joseph Larmor, would have felt dishonored had one of their distinguished colleagues declared himself a conscientious objector. Trying to convince Eddington or to implicate him was out of the question. There were attempts to intercede with the Home Office on behalf of Eddington on the grounds that it was not in England's best interest to send such a distinguished scientist to the front. The death of Harry Moseley in August 1915 at the Turkish front was still fresh on everyone's mind. Larmor and the others came close to seeing their efforts rewarded.

But Eddington had his own idea. When asked to sign an ad hoc letter, he added a postscript stating that even if not drafted, he would request conscientious objector status anyway. Would he be sent to "peel potatoes" in some camp? He might well have been, like some other Quakers. But Dyson's plea to the Admiralty permitted a diplomatic resolution of the problem. Eddington obtained a deferment of military duty on the express condition that, if in 1919 the war were over, he would head one of the expeditions.

They sailed from Liverpool on 8 March 1919 for a trip that would last six months. In Madeira, the two expeditions went their respective ways: Eddington and Cottingham to Principe and Davidson and Crommelin toward the mouth of the Amazon, which they reached on 23 March. Morize, the director of the Rio Observatory, who was supposed to receive them, was not at Sobral, so the astronomers spent a month in Manaos. They were ready to get down to work on 30 April. In Dyson's report, the Brazilian delegation is described in detail. There were representatives of the local authorities, civilian and ecclesiastical: the police chief, the bishop, and the district congressman, who had lent his house, across from the Sobral Jockey Club racetrack, where the astronomers set up their equipment. On 9 May, Morize joined them, together with the Brazilian expedition that would carry out spectroscopic observations of the solar corona. Soon, everything was ready for the D-day everybody had been anxiously anticipating.

"The eclipse day opened very unpromisingly, the proportion of cloud at first contact being about 9/10. . . . A large clear space in the clouds reached the Sun's neighbourhood just in time, and four out of the five minutes of totality the sky round the Sun was quite clear."[17]

In Africa, at Principe Island, Eddington and Cottingham were not so lucky. The morning of the eclipse, they feared a storm, and the pictures were taken through persistent clouds. From Sobral, Crommelin cabled London: "Eclipse splendid," whereas Eddington, despite his optimistic disposition, sent an unenthusiastic "Through cloud. Hopeful." The photographs were developed. Only two shots from the astrograph at Principe were satisfactory, and they showed only six or seven stars. At Sobral, the team was delighted: the pictures taken with the four-inch telescope showing seven stars were considered satisfactory to excellent. Those from the astrograph, on the other hand, were disappointing, for the stars are somewhat diffuse. The focus of the astrograph appears to have been deficient, probably due to the distortion of the image on the coelostat's mirror, combined with the drop in temperature, even if small, at the time of the eclipse.

They all packed up and left. Davidson and Crommelin returned to Sobral a few days later to take the comparison shots (of the same star field but without the Sun) on site. Since the photographs taken with the four-inch telescope were too big to be measured with the micrometer at the Royal Observatory, they decided to invert the comparison plates so that the measurements could be made "by superposing" the photographs—a novel method. For practical reasons, the comparison plates of the Principe team had to be taken on their return to Greenwich, which may have created some additional problems related to the calibration of the instruments.

The final article was written by Dyson, Eddington, and Davidson. Oddly enough, Cottingham and Crommelin are not among the authors, and the names of the authors are not in alphabetical order. Moreover, Dyson's name appears first, although he did not take part in the expedition, but he was the Astronomer Royal and spared no effort in making it happen. Eddington had headed the Principe expedition, even if he did not report the most interesting results; despite being the youngest, he was the most actively involved and the true representative of English astronomy

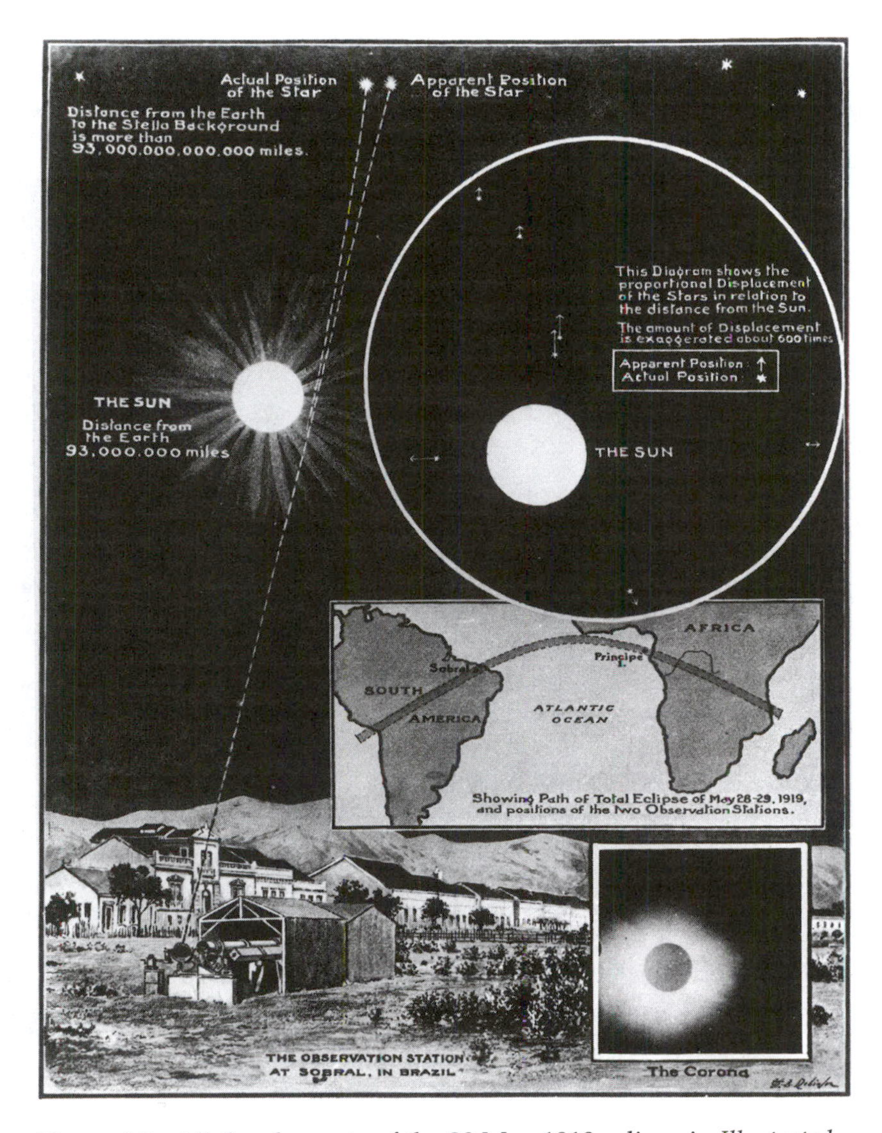

Figure 8.5. All the elements of the 29 May 1919 eclipse in *Illustrated London News*. (© *L'Illustration*/Sygma)

in this enterprise. As for Davidson, he had headed the Sobral expedition, but Eddington and Dyson were in charge of the analysis and interpretation of the data. The measurements were performed at Cambridge, and there were some very tricky interpretation problems.

Figure 8.6. The total solar eclipse observed by a French mission at Poulo-Condore. (© *L'Illustration*/Keystone)

Recall that the possible results had been quantified by Eddington: no deflection at all, a 0.87″ deflection, or a "double" one of 1.75″. The first two outcomes would support Newton's theory, whereas the third one, the one Eddington yearned for, would be the deflection predicted by Einstein's theory of gravitation.

But the fact that the deflection effect decreased rapidly with the distance of the star to the center of the Sun had to be taken into consideration. There were other problems as well: the shifting and rotation of the photographs, which may have been be minimal but had to be subtracted to obtain the deflection really due to the Einstein effect; the fact that the comparison shots were not taken at the same time—and, in some cases, not at the same place; additional effects (aberration and refraction) that required a correction from one photograph to another; and finally, so-called systematic errors that also had to be corrected. In short, all that will have to be carefully analyzed, and will open the door to complex, difficult, and impassioned discussions.

After all the analyses and corrections were made, the measurements from the four-inch telescope, which were considered to be the best, showed a mean deflection of 1.98″ plus or minus 0.12, and those from the Principe astrograph, a mean value of 1.61″ plus or minus 0.30, the larger error interval being due mainly to the number of visible stars; the more and better distributed those are, the more precise the results.

As for the plates from the Sobral astrograph, they showed 0.92″ (close to the Newtonian value), but their quality was such that no error estimate could be given. It was decided to disregard those results and take the Principe ones into account, assigning the heaviest weight to the results from the four-inch telescope, the most favorable to general relativity. This question of the relative weight of observations (which can hardly be objective, for it ultimately depends on the global idea that an astronomer has of his own work) was in fact extremely important. And it would be bitterly debated, due to the long-held suspicion that Eddington and Dyson, who were responsible for the interpretation of the data, had favored the measurements that best agreed with the hoped-for results: Einstein's.

All this explains why, while impatiently waiting for results that were thought to be significant, the announcement of the Goldengale preliminary figures in July 1919 was received with caution. Campbell nevertheless thought he could claim, as we have seen, that Curtis's results refuted Einstein's theory, for they could only be interpreted as consistent with Newton's theory, which predicted a deflection half as large as that derived from general relativity. In fact, due to the difficulty of accurately measuring from the plates and various other complications, the results from the Goldengale eclipse were never published.

"GREEK DRAMA"

Both at the Joint Eclipse Meeting in London, on 6 November 1919, and in their final article, Eddington and Dyson left no room for doubt: "After a careful study of the plates I am prepared to say" affirmed Dyson, who spoke for the Sobral team, "that there can be no doubt that they confirm Einstein's prediction. A very definite result has been obtained that light is deflected in accordance with Einstein's law of gravitation."[18] As for Eddington, whose conclusions on the Principe observations were much less clear, he supported Sobral's results: "This result supports the figures obtained at Sobral."[19]

Such was also the conclusion of the article, which emphasized the highly exceptional conditions of the 1919 eclipse: "But the observation is of such interest that it will probably be considered desirable to repeat it at future eclipses. The unusually favourable conditions of the 1919 eclipse will not recur, and it will be necessary to photograph fainter stars, and these will probably be at a greater distance from the sun."[20] The future would bear out this analysis.

Under the chairmanship of Sir Joseph John Thomson (the discoverer of the electron), both the Royal Society and the Royal Astronomical Society assembled at the special joint meeting of 6 November 1919, which was devoted to the results of English expeditions. Alfred Whitehead gave a gripping and somewhat embellished account of the meeting, where the "dramatic triumph" of general relativity was declared: "The whole atmosphere of tense interest was exactly that of the Greek drama: we were the chorus commenting on the decree of destiny as disclosed in the development of a supreme incident. There was dramatic quality in the very staging: the traditional ceremonial, and in the background the picture of Newton to remind us that the greatest of scientific generalizations was now, after more than two centuries, to receive its first modification. Nor was the personal interest wanting: a great adventure in thought had at length come safe to shore."[21]

It was a remarkable day, although a difficult one for the British scientists who, in a way, lost their preeminent place in the pantheon of science. After having dominated the queen of sciences for more than two centuries, their hero was dethroned. But at that highly symbolic moment, Whitehead, anxious to play fair, timidly spoke of a "first modification." He was reluctant to endorse an outright refutation of Newton's theory; the term "modification" suggests that, on the whole, the theory remained valid, which was indefensible, both from a theoretical and an epistemological point of view. And yet, from a practical perspective, Whitehead was not completely wrong, for Newton's theory, of which there was *theoretically* nothing left as far principles were concerned, would continue to dominate gravitation: the calculations, the observations, and the scientists themselves, almost until the 1960s.

The meeting was not as smooth and ceremonious as Whitehead suggested. After the astronomers had presented the results of the expeditions and their interpretation, Thomson opened the discussion with a few comments. He first recalled that in *Opticks*, Newton had suggested that light was subject to gravitation but that his hypothesis produced only one-half of the deflection that had been predicted by general relativity and now had been measured. He then observed that "this is the most important result obtained in connection with the theory of gravitation since Newton's day." Thomson then praised "not so much his [Einstein's] results" as "the method by which he gets them," before adding that "the weak point in the theory is the great difficult in expressing it. It would seem that no one can understand the new law of gravitation without a thorough knowledge of the theory of invariants and of the calculus of variations."[22] Shortly afterward, "a mere physicist" spoke to drive the point home, regretting that the theory "is always presented from a purely mathematical point of view" and refusing to believe "that a profound physical truth cannot be clothed in simpler language."[23]

Next, Ludwik Silberstein spoke at length to plead for caution. He ended his long-winded speech by pointing to Newton's portrait at the back of the room and declaring, not without grandiloquence: "We owe it to that great man to proceed very carefully in modifying or retouching his Law of Gravitation," and he added, in a low voice "This is by no means defending blind conservatism."[24]

The results of the 1919 eclipse were of great significance for general relativity and marked a turning point for the theory. In particular, for certain experts who had not been convinced by the explanation of Mercury's perihelion, the 1919 results constituted the deciding factor in favor of general relativity. The 1919 results therefore made their mark, all the more so given that for almost forty years there would be nothing else apart from these results to support general relativity; results that, little by little, would lose their strength in view of the failure of other attempts at experimental verification. But the 1919 eclipse also marked a turning point for Einstein's image, whose glory until then came more from high society than from scientific circles.

In 1922, a Lick expedition headed by Campbell confirmed the 1919 results. The reaction of the American astronomical community was mixed. There were several enthusiastic responses, but many scientists were lukewarm, especially among the astronomers. Perrine could not accept the theory's conceptual framework: "The whole relativity business has seemed to me unreal and so purely philosophical that to accept it is to upset our previously carefully constructed and very material systems. . . . I am open minded but conservative in this matter."[25]

Curtis, too, was unable to accept the latest results: "There may be a deflection, but I do not feel that I shall be ready to swallow the Einstein theory for a long time to come, if ever. I'm a heretic."[26] In a letter to a correspondent, he confessed: "I have never been able to accept Einstein's theory. This in spite of the fact that many eminent mathematicians regard it as the greatest advance since Newton's time. I regard it as a beautifully worked out alternative 'reference frame,' apparently adequate, but by no means essential and by no means necessarily the correct system of reference. . . . We do not force ourselves to accept non-Euclidean geometry simply because it seems to 'fit.' Perhaps I am wrong, but it does not seem to me at present that I shall ever be willing to accept Einstein's theory, beautiful but bizarre, clever but not a true representation of the physical universe."[27]

Lenard had recently reprinted in *Annalen der Physik* part of the original Soldner article, his goal clearly being to discredit Einstein by insinuating that he had essentially copied Soldner. Some American newspapers carried the story of Einstein's apparent deceit to make him look like a "humbugger," a "fraud."[28]

Following a request from Campbell, Robert Trumpler, one of the rare American astronomers specializing in general relativity, wrote a historical note in whose conclusion he asserted "the independence of Einstein's work."[29] Trumpler (correctly) noted that Einstein was probably not aware of Soldner's results, which had been completely forgotten since the development of the wave theory of light, at the turn of the nineteenth century. We know today that Einstein was not familiar with those results, but that did not prevent him from thinking "that the idea of a bending of light appeared at the time of the emission theory is rather natural," as he wrote to Freundlich in 1913.[30] Knowledge of Soldner's results, however, would not have been of any help in the development of his theory, for it was an entirely different logic that led him to general relativity, and the bending of light rays was only a consequence of it, one which would not have come as a surprise to him.

The history of the development of general relativity and of relativistic techniques would remain far removed from the logic of light for quite a while. We shall come back to this fact—a rather regretful one—for light is an object, a concept, an element essential to general relativity and the study of its properties. As I shall show in chapter 12, it is precisely because light did not have its proper place in the theory that the idea of a black hole, whose fundamental properties are linked to light, would not appear until the 1960s.

SUMATRA 1929

Other verifications of the deflection effect regularly took place. Campbell, disappointed at the failures of the Lick Observatory, at Brovary, but especially at Goldengale, prepared an expedition to observe the eclipse that crossed Australia on 21 September 1922. The reported results roughly confirmed the 1919 observations. Freundlich, who, let us recall, had studied this effect since 1911, was also on the path of the Australian eclipse, but the weather was once again uncooperative. Luck was not on his side. He had been betrayed by rain, imprisoned by the Russians, and overtaken by the English and the Americans. Extremely eager then, after the tremendous success of 1919, to make his own contribution, he organized first in Mexico, in 1923, and then in Sumatra in 1926, expeditions that rain once again upset. Finally, in 1929 in Sumatra, he foiled his curse, but he was embittered by his repeated failures.

Learning from his own setbacks and from those of other expeditions, which he had analyzed in detail, Freundlich had taken all possible precautions as far as the weather and the instruments were concerned. But he could not choose the eclipse. The results from Sumatra were markedly different from those he had expected and would puzzle both practicing astronomers and theoreticians: 2.24″ (plus or minus 0.10″). This was too strong a deflection, and it tended to refute Einstein's theory. The astronomers admitted as much in the conclusion of the article in which they reported their results two years later: "There seems to be no doubt possible, therefore, that our series of measurements does not agree with the value 1.75″ claimed by the theory."[31]

In fact, Freundlich's case hinged on how the probable error of the measurements was calculated: if the margin of this error was relaxed, his results were then consistent with relativity. His 0.10″ probable error was vigorously challenged, because he emphatically claimed that his result called into question those of 1919–1922 and therefore the very foundations of the observational support for general relativity.

In 1931, Freundlich communicated his results to the Royal Astronomical Society. His conclusions may be summarized in three points: "(1) A deflection exists. (2) It is not Newton's. (3) It seems to be greater than Einstein's."[32]

Present at the meeting, Eddington found it "difficult to believe that 1.75″ can be wrong." However, his confidence wavered, for he said: "Light is a strange thing, and we must recognise that we do not know as much about it as we thought we did in 1919; but I should be very surprised if it is as strange as all that."[33]

Not everybody remained as calm as Eddington, and doubts seized some relativists, including Einstein himself. He reacted to the results of the 1929 eclipse in a philosophical way. He was then, and had been since the 1920s, way beyond his general relativity, which he now considered to be only the basis, a support, upon which he sought to build a new theory, a unified theory of relativity and electromagnetism. Completely taken with that work, Einstein did not appear to be bothered by those anomalies, for his new theory could predict quantitatively different values for the phenomenon.

In early 1930, Paul Langevin's conclusions were decidedly cautious, coming from someone who could not be accused of being antirelativist: "The deflection of light by the Sun, quantitatively calculated by Einstein, was experimentally verified during the 1919, 1922 and May 1929 solar eclipses. The results are extremely difficult to evaluate in the sense that we are dealing with measurements at the limit of instrument precision, especially regarding the stars that are somewhat distant from the Sun in the direction."[34]

At the first international conference on general relativity held in 1955 in Bern, it was Trumpler (now the most respected expert on the subject) who was invited to talk on the present state of the observations. The ambiguity of his conclusions mirrors the prevailing sentiment: "If one considers the various instruments and methods used and the many observers involved, the conclusion seems justified that the observations on the whole confirm the theory."[35]

Such was indeed, "on the whole," the general view of the experts, with the exception of Freundlich. A view that raised, one more time, the question of the validation of a scientific theory. It is a process that never really ends, for there always comes a time when a theory will be replaced by the newly arrived one whose results are more precise.

At the beginning of the 1950s, we find an echo of this debate in a correspondence between Born and Einstein. Born attended a talk given by Freundlich, who had become more aggressive with the passage of time, and he conveyed his uneasiness to Einstein, who replied that "Freundlich . . . does not move me in the slightest."[36] A calm that in the circumstances was perfectly justified.

And yet, in those postwar years, the results disputed by Freundlich did not help the cause of general relativity, which was not in very good shape, but not because something new had come up. Everything was going reasonably well, but the contact between the theory and reality was terribly weak. Except in cosmology, where it had somewhat increased its role, general relativity had the greatest difficulty in producing new results, finding new applications, and conquering new followers. And most theoretical physicists preferred to work on the quantum theories that were infinitely more productive in terms of results.

NOTES

1. Soldner, 1801.
2. Cf. chapter 12 and Eisenstaedt, 1991, 1997.
3. In this regard, cf. Einstein to E. Freundlich, August 1913, *CPAE*, vol. 5, p. 550, and Eisenstaedt, 1991, p. 378.
4. Einstein, 1911, pp. 107–108.
5. Einstein to E. Freundlich, 1 September 1911, *CPAE*, vol. 5, p. 317; English translation, vol. 5, pp. 201–202.
6. Ibid.
7. Light rays are refracted by the Earth's atmosphere as if they were entering a crystal. This atmospheric refraction is a function of the altitude over the horizon of the observed phenomenon: the lower the observation, the stronger the refraction, thus affecting the quality of the photograph.
8. Crelinsten, 1984, p. 20.
9. G. E. Hale to H. D. Curtis, 24 June 1919; quoted in Crelinsten, 1984, p. 42.
10. Curtis, 1917, p. 63.
11. Ibid., p. 64; quoted in Crelinsten, 1984, p. 27.
12. W. W. Campbell, 1919, in Royal Astronomical Society, 1919, p. 299.
13. Eddington (no date); quoted in Chandrasekhar, 1969, p. 578.
14. Quoted in Crelinsten, 1984, p. 25.
15. See, 1916, p. 512.
16. Jeans, 1917, p. 57; quoted in Crelinsten, 1984, p. 25.
17. Crommelin, 1919a, p. 370.
18. F. W. Dyson; quoted in Royal Society and the Royal Astronomical Society, 1919, p. 391.

19. A. Eddington, 1919; quoted in Royal Society and the Royal Astronomical Society, 1919, p. 392.

20. Dyson et al., 1920, p. 332.

21. Frank, 1950, pp. 217–218.

22. J. J. Thomson, in Royal Society and the Royal Astronomical Society, 1919, p. 394.

23. Quoted in Royal Society and the Royal Astronomical Society, 1919, pp. 394–395.

24. L. Silberstein, in Royal Society and the Royal Astronomical Society, 1919, p. 397.

25. C. D. Perrine to W. W. Campbell, 1923; quoted in Crelinsten, 1984, p. 81.

26. H. D. Curtis to Chant, 1923; quoted in Crelinsten, 1984, p. 81, and Earman and Glymour, 1980a, p. 68.

27. Curtis to Dr. Vogtherr, 19 September 1923; quoted in Crelinsten, 1984, p. 81.

28. Quoted in Crelinsten, 1984, pp. 82–83. On Soldner's article and this grim Lenard affair, cf. Jaki, 1978.

29. Trumpler, 1923.

30. Einstein to Freundlich, August 1913, German original in *CPAE*, vol. 5, p. 550; English translation, vol. 5, p. 351.

31. E. Freundlich, 1931, quoted by Hentschel, 1994, p. 181.

32. E. Freundlich, in Royal Astronomical Society, 1932, p. 4.

33. Royal Astronomical Society, 1932, p. 5.

34. Langevin, 1932, p. 227.

35. Trumpler, 1956, in Mercier and Kervaire, 1956, p. 108.

36. Einstein to M. Born, 12 May 1952, in Born, 1969, p. 206 of the English translation.

CHAPTER NINE

Relativity *Verified*: The Line Shift

THE REPUTATION OF general relativity as a difficult theory has often been blamed on the sophisticated mathematical apparatus that its precise formulation requires, and this is certainly the main reason for the vague uneasiness it provokes. But isn't the tree hiding the forest? It is not clear that we are looking in the right place: just as in special relativity, the crux of the problem is above all conceptual, as the question of the third test will show.

The principle of equivalence is like a small theoretical machine that allows us to replace a gravitational field with an acceleration, to *handle* gravitation with the concept of acceleration. We can solve in this way a variety of problems, the simplest one being the behavior of clocks (clocks, of course! one of relativity's most important and meaningful concepts!) in a gravitational—and hence, an acceleration—field. What happens if we set a clock in motion, if we accelerate it, and later compare it with its little sister left behind on Earth?

What happens when a clock is accelerated or—which amounts to the same thing, by the principle of equivalence—subject to a gravitational field? We could compare the time ticked away by two clocks, one of them constantly accelerated (for example, by a constant gravitational field) while the other, its twin, would remain near the observer on Earth. We will then be able to compare time in two different gravitational fields, the Earth's and the Sun's, for instance. Information will be transmitted in the form of electromagnetic signals. The first signal will carry the 12:00 beep, and the second one the 12:01 beep. On reception of these two signals some eight minutes later, the time it takes for them to travel back to Earth, we will find out how long a solar minute is compared to the terrestrial minute marked by the twin clock on my table.

This is an almost perfect experimental protocol—and a wholly relativistic experiment.

Einstein observed that such clocks exist everywhere in the universe and that their rate can be monitored. They are the atoms, whose spectral lines are small clocks. This is not surprising if we recall that (*proper*) time may be very accurately defined here on Earth by the ticking of a cesium clock. In a gravitational field all these clocks will be slowed down, and the lines of our atomic spectrum will be shifted toward the red. The reasoning that led Einstein to this conclusion has to do with the equivalence between a gravitational field and an accelerated reference frame. It suffices for us to replace the first with the second, and we should arrive, if we do not make any mistakes, at the following result: the shift is directly proportional to the difference in strength of the gravitational fields at each of the two locations.

The experimental protocol is extremely simple: it is enough to observe a particular emission line, from a sodium atom, for example, on the Sun and down here. The photons emitted by the atom on the Sun should reach the observer on Earth, which is subject to a different gravitational field, with a different color from the one they had at the time of emission; their frequency will be shifted. This shift of the particular line in a gravitational field is also called the Einstein effect. The physical effect is therefore clearly identified, but two things remain to be done: to find the formula to calculate the effect and to determine whether the effect actually takes place.

DEFINING THE EINSTEIN EFFECT

The definition of the Einstein effect will be meticulous, bordering on finicky, but this is not surprising if we take a closer look. Each of our two observers, denoted O_1 and O_2, will study the emission spectrum of two atoms, A_1 and A_2. Let us consider two identical experiments, one performed on Earth (O_1 observes A_1) and another on the Sun (O_2 observes A_2). Two identical atoms emitting identical lines are observed by nearby observers using identical spectrographs; two strictly local experiments that should produce the same result: the frequency of atom 1 at point 1 is therefore equal to that of atom 2 at point 2: $\nu_{1,1} = \nu_{2,2}$. This is the first fact. Let us note by the way that we have applied here the principle of general relativity: physics is the same (the answers are identical) regardless of the site of the experiment, which is not entirely obvious given that the gravitational fields are different.

Now, O_1 wants to observe the other atom, A_2, located at a distant place and in a different physical context. More precisely, O_1 wants to measure the frequency of the same line of this (identical) atom subject to a different gravitational field, and finally compare it with the local measure (O_1 observes A_1). It is then a question of comparing the proper frequency $\nu_{1,1} = \nu_{2,2} = \nu_0$, with the frequency of atom A_2 observed by O_1 in 1: $\nu_{1,2}$. The algebra is not difficult but the calculation requires a delicate analysis.

Everything is (relatively!) simple on the surface, but to carry out the calculation accurately and not get lost along the way, we need to consider four trajectories (just as for the perihelion in chapter 7). First of all, one for each atom, assumed to be at rest in different gravitational fields (we must describe their motion even if they are at rest—"rest" being a relative concept); then, the trajectories of each electromagnetic signal (i.e., those of the photons) carrying the information, that is, the "beeps" allowing us to define the period, corresponding for example to the peak value of the signal. These trajectories must be written in a particular coordinate system, and we must be careful not to confuse coordinate time and proper time, an elementary precaution in general relativity, although not always strictly observed during the first half of the twentieth century. The final formula is simply written:

$$\frac{\nu_{1,2} - \nu_{1,1}}{\nu_{1,1}} \propto \frac{GM}{rc^2}$$

The frequency variation is proportional to the gravitational potential to which the atom is subjected, GM/rc^2. It is not a difficult exercise, if one is careful and does not confuse proper time and coordinate time, distinguishes the two observers and the two atoms, and does not get lost in the indexing. These are the things that a rigorous theoretician would do. Still, it is necessary to tackle the question from the right angle.

But aren't we forgetting some tricky questions? We have assumed that proper frequency is invariant. That much is obvious; but we have also assumed that the frequency (of an atom line) is an invariant, and this is a hypothesis. Einstein assumed that the frequency of each spectral line of any given atom is the proper frequency. If it is measured by a nearby observer, the result will always be the same, regardless of the gravitational field common to both. The yellow emission line of sodium measured by an observer at rest with respect to the atom will always be 5,893 angstroms. It is a hypothesis that must and does hold, directly or indirectly. But we must emphasize that the hypothesis according to which the frequencies of atomic spectral lines are invariant is not an innocent one: it is central to the theory and central to physics; a hypothesis on which every experiment involving frequencies is necessarily based.

I hope the reader has appreciated that the proof of the formula giving the Einstein effect, the third test of general relativity, is not, technically speaking, very complex. But it is *subtle*. More than one of our distinguished experts, specialists, and professors have been deceived by it. I will spare the reader the list of reasoning, calculation, or conceptual mistakes that appear in the literature in connection with this proof. In order to convince those still skeptical about such mistakes, here is an anecdote as told by Hendrik Casimir, who had Wolfgang Pauli as a teacher. Pauli was the author of a brilliant book, one of the first on the theory: "Yet I remember that once when he was speaking about the so-called red shift . . . , he obtained an expression with the wrong sign, which meant a shift towards the violet instead of towards the red. He then began to walk up and down in front of the blackboard, mumbling to himself, wiping out a plus sign and replacing it by a minus sign, changing it back into a plus sign, and so on. This went on for quite some time until he finally turned to the audience again and said: 'I hope that all of you have now clearly seen that it is indeed a red shift.' "[1]

What student has not witnessed a similar scene, what teacher has not suffered a comparable agony? It may happen when the teacher has not properly prepared his or her course; it can happen to everybody, including the best ones. Only, in general relativity it happens more often. And it makes you wonder. If it occurs during a lecture, it can be forgiven, but why should so many mistakes of that kind appear in textbooks? Mistakes that are not simply due to a wrong sign but often to a wrong understanding or formulation of the problem, such as confusing coordinate time and proper time or one atom with the other. But, as Pauli did, we land on our feet and sometimes find, miraculously, the formula. All the great of this world were deceived by the simplicity of the proof: Einstein the first, von Laue, whom I quoted in the introduction, Hermann Weyl, one of the most serious authors, and Eddington himself.

What should we make of this? That our best specialists are not very serious? It would be an amusing conclusion but not a convincing one. That they do not master their subject? That would be going too far. But then, what is going on?

Actually, there are two kinds of difficulties. To begin with, conceptual difficulties. We must be careful in stating the problem, take the time to analyze it correctly, and not confuse the points of view or the coordinates. That is where our authorities usually get lost. Second, even if it is often only a simple rule of three, we must accept that, despite the apparent simplicity of the question, we are dealing with a complex problem that cannot be solved without proper precautions.

In this respect, our distinguished professors may suffer from a bit of self-importance. Faced with a seemingly very simple problem, they want to go too fast; they believe it can be solved in no time and wish to get it over with. Besides, don't they all (or almost all) finally arrive at the right formula? The tune is well known and, it is that of Pauli's "I hope that all of you have now clearly . . . ."

The experts would certainly be more at ease dealing with a

more technical problem that requires a development several pages long and where algebra would take them safely to the destination. Once the problem is correctly stated, it would be enough to follow the calculations step by step. This takes us back to our initial point and to what is one of the major difficulties of the theory: to choose the right concepts, fully grasp them, clearly understand what is invariant, and not be misled by the relativity of the chosen formulation or by the coordinate system, to which we mechanically attach an immediate physical interpretation. To conceive space-time is not so simple.

Which brings us to problems involving twins since, just as our atoms, our clocks are also identical twins. The twins' paradox is a striking and well-known example of the difficulty to think about and to accept relativity (special as well as general), an example much written about and long questioned by reputable experts. The problem perfectly mirrors our experiment with atoms. Two twins leave from the same place but in different vessels. One of them travels by rocket and after a certain time rejoins his brother. Given the homogeneous and isotropic character of the special relativity space in which these journeys take place, how is it possible that one of the twins should age more than the other if they are going to be at the same place at some later time? Isn't the situation symmetric? It only *seems* symmetric. Why, then, should one of the twins age more than the other? And why not the other more than the one, if I may put it that way? It is because their journeys are not identical, and they can end up at the same place without the *proper durations* of their trips being the same. They have traveled along different paths in space-time; each of them was accompanied by his proper time that was recorded by "the clock that he carries in his hand": their ages are not the same on arrival. Here, too, there are many *times*: coordinate times and proper times, the only ones to have a physical meaning. Experiments of this type have been actually performed thanks to identical clocks placed aboard airplanes or satellites and have confirmed the phenomenon of the slowing down of clocks.

Relativity, whether special or general, supposes the comparison of (at least) two situations—the platform and the train, the

Sun and the Earth—and we must be able to go from one to the other without becoming dizzy or getting lost. It is not that easy, and the history of relativity shows that everybody, from students or the general public to the experts and occasionally even Einstein himself, may lose their way. It is the case of the twins' paradox—so called for a reason—or even the proof of the third test, where similar difficulties arise. These difficulties are not technical (the *proofs* of the Einstein effect are not complex from a mathematical point of view) but conceptual: of seeing the problem in the right way, of formulating it correctly, and of mastering the novelty of the concepts involved.

Relativity forces the physicist, in the course of his or her argument, to go from one system to another, to change the point of view, and it is not so easy to stay the course and not get lost. In relativity's space-time (be it special or general) there is no absolute time, and we must accept that only physical time, that is, proper time, is a function of the actual trajectory. Of course nothing like that can occur in Newton's theory, where God keeps track of a time that is the same for everybody.

In fact, we are all familiar with that kind of problem: Isn't it difficult to imagine the motion of the Earth around the Sun? To *see* the phenomenon of the duration of the day or the seasons is not an easy task, and it takes a particular choice of the point of view, geocentric or heliocentric, as the case may be. But, unlike what happens in a Newtonian universe, in general relativity we can no longer count on God to watch over that absolute space in which each object and each observer has a well-defined position, motion, and time. The objects seem to float in a somewhat unreal space: they are in a kind of free fall. As I put it in a rather crude fashion in chapter 5, it is a matter of going from one train to the other while keeping an eye on the stationmaster standing on the platform. It is not surprising then to find that this kind of reasoning is particularly subtle: we must go from one coordinate system to another, come back to the first one, etc., all those systems being very similar. We have to struggle with a relative environment and a lack of an absolute and invariant reference point. This may cause a feeling of bewilderment and which we deal with in different ways: perhaps by avoiding the problem or by solving it in a superficial way, in short, by refusing to confront it. It is the flight, a feeling of urgency that propels us toward the exit, the formula: let's move on, all that is well known; this is probably what happened to Pauli. Another way of coping may be by retreating to an almost-Newtonian point of view, one of the first-generation relativists' favorite solutions. They entertained a neo-Newtonian vision of their theory, in a sense rejecting the very essence of that theory. And still another approach would be to confront those questions in their relative complexity. This requires rather heavy artillery and multiple precautions, and a calmer, simpler approach to the problems. This is why it is important to identify each observer, each atom, and each trajectory carefully.

We must not forget the poor students who had to learn relativity in such a quagmire. For physics is not mathematics—not only—and rigor comes only gradually and slowly. Cleary defined practices are put into place little by little, and in the meantime it is not unusual to see some fuzziness reign. But this does not mean that the theory is wrong.

Meanwhile, the first generations swim, float, or sink. It takes a good dose of flexibility to stand fast in front of a theory whose presentation and interpretation may at first appear contradictory and paradoxical. And the best specialists are not the most rigorous. To correctly practice this profession one must (also) know how to clear obstacles and ignore certain imprecisions, which does not mean to permanently accept them. Einstein was not reluctant to take daring conceptual and technical leaps—some of them extremely questionable—(and leave the proofs to his colleagues) if not for a good reason, at least for a reason he then thought to be right, for it often happens that a promising interpretation goes hand in hand with a certain fuzziness of the conceptual framework and even with some rather doubtful mathematical practices.

For these reasons, should we remain totally uncompromising and not go forward except when absolutely sure? Certainly not; if we wish to make progress, we must be willing to accept the vagueness of certain ideas. Being too demanding would condemn us to immobility. The numerous errors and inaccuracies we now see did not stop our specialists, for they expected to be able to fix them eventually. A physical theory does not come ready-made like Minerva from Jupiter's head. It is built little by little with the contributions of many people. As Kuhn put it in his *The Structure of Scientific Revolutions*: "Mopping-up operations are what engage most scientists throughout their careers. They constitute what I am here calling normal science."[2] And it is thanks to that kind of work that the theory progresses, that it is simplified and "recast." And so it gradually becomes more accessible as more consistent, clearer, and simpler tools, practices, and interpretations appear. Our experts had no doubt better things to do and preferred a "good" heuristics (almost by necessity ambiguous) to a proof several pages long where the physics would have been concealed under a rather stifling formalism. But we can understand that, when confronted with those fuzzy proofs, the frailest students go to pieces.

Those proofs have two essential functions. The first one is heuristic: to produce understanding and to *show* how the theory works, "to get a feeling" for it; the second one is to write out the laws of the proof (which is done despite the sometimes repeated mistakes) and so to ensure the mastering of the theory on the logical level. Now, the most important feature for a physics text is the heuristics. Einstein does it masterfully, and the more points of view the better.

As for the second function of those proofs or *demonstrations*, its purpose is to guarantee that all the logical, mathematical, formal, and conceptual arguments are impeccable and may be developed from A to Z without flaws—in short, that the concepts are clear, well organized, and related to equally clear variables or observables; that the connection between the mathematical and physical levels is soundly established (which is not always easy). Once we are sure (as sure as we can possibly be) that the conceptual structure of the theory is faultless, we must verify that the mathematical relationships between the variables representing the physical concepts are perfectly integrated and also that the mathematical framework of the theory is consistent. Such a recasting of the theory generally requires a lot of energy and time—and it takes up a lot of space in the discipline's literature—a process that may last for decades. At the present time, this is practically achieved. But if we want to understand, we must get down to business right away and make sure that the physical effect is really there.

THE OBSERVATION OF THE THIRD TEST

Early on, Einstein came up with the idea of comparing the emission frequencies of an atom's spectral lines on the Sun with those on the Earth. A shift of the spectral lines on the Sun had been observed before the beginning of the twentieth century, but the reason for it was not really known. It was not an easy phenomenon to analyze, particularly in that domain, and the German astrophysicists as well as the French spectroscopists generally doubted that they were in the presence of the effect predicted by Einstein. It was far from certain that the slight shift they observed was due to gravitation, for there was no shortage of other reasons to explain it; reasons related to the physics of the Sun, which was only partially understood through the study of the pressure, the dispersion, and the width of the lines.

On the question of the red shift, Einstein wrote to a physicist at the University of Utrecht who claimed that the observed red shift of the Sun's spectral lines may be due to a dispersion phenomenon. If such was the case, Einstein admitted, "then my cherished theory must go in the wastepaper basket."[3]

This was only the beginning of a long and complex story. The least that can be said is that the verification of the Einstein effect on the Sun was not an open-and-shut case. Between 1915 and 1919, numerous measurements were performed, generating results whose interpretation was largely negative; but from 1919 on, measurements made on the Sun by German spectroscopists seemed to go in Einstein's favor. Shortly afterward, the French spectroscopists also reported interesting results. The tide was turning. And we can wonder, as people did then, if the results of the 1919 eclipse had something to do with it. Since general relativity had been verified and accepted, it was only natural that the shift effect should exist and was observed. Einstein would remain forever optimistic with respect to the red shift. He believed in it. In December 1919, he wrote to Eddington: "According to my persuasion, the red-shift of the spectral lines is an absolutely compelling consequence of relativity theory. If it should prove that this effect does not exist in nature, then the entire theory would have to be abandoned."[4]

This was also Silberstein's view, although he argued in the opposite direction: the eclipse's result was doubtful because the shift had not been observed. During the November 1919 joint meeting of the Royal Society and the Royal Astronomical Society, he declared with his characteristic grandiloquence: "It is unscientific to assert for the moment that the deflection, the reality of which I admit, is due to gravitation. . . . The discovery made at the eclipse expedition, beautiful though it is, does not, in these circumstances, prove Einstein's theory." Then, after having paid a vibrant tribute to Newton, he came to the third test: "If the shift remains unproved as at present, the whole theory collapses"[5]—a collapse that he would have liked to see happen and was gloatingly waiting for. It would be naïve to believe that Einstein's success was universally acclaimed. Certainly not. After all, these were men, as vulnerable to an occasional fit of jealousy as anybody else. And besides, isn't sometimes the fall of the great a source of secret pleasure to others? A fall that is most spectacular when the tightrope walker has set the rope defiantly high; a fall announcing that the place is now clear for your own ideas. And anyway, there are so many reasons to reject this awfully complex theory!

But let us go back to the shift. It is clear that even if the proofs of the formula were not perfect, the existence and consistency of the Einstein effect were accepted by everyone, and it is generally agreed that the survival of the theory depended on that acceptance.

The spectral shift, just as the two other classical effects of general relativity, is of the order $2GM/rc^2$. Hence, at the edge of the Sun, the frequency variation or *shift* due to the effect should be of order 10^{-6}. This is a very small effect compared with other expected or possible ones, and it would be desirable to find conditions resulting in a larger order: on a star more massive or of smaller radius than the Sun, for instance. This is precisely what happened in 1925. A new testing ground for the Einstein effect opened up: white dwarfs.

The characteristics of the companion of Sirius, a star that revolves around Sirius and forms with it a double-star system, were then fairly well known. Thanks to various orbital and spectroscopic observations, it was believed that the companion of Sirius had a mass nearly as large as the Sun's and a radius smaller than that of Uranus, hence an enormous density, of the order of 60,000 g/cm^3. It would take a long time to understand how such a star could be in equilibrium. (These exotic and peculiar stellar objects were just then being observed for the first time, and it is a fundamental chapter in the history of astrophysics that begins: *relativistic astrophysics*, to which we shall return in chapter 15.) The companion of Sirius gravitational potential ($2GM/rc^2$) was therefore thirty times that of the Sun's,[6] and this provided an opportunity to verify the Einstein effect in better conditions. In 1925, at Eddington's request, an astronomer at Mount Wilson Observatory detected a shift of precisely the order of magnitude of the effect predicted by the theory. Eddington was exultant: "Professor Adams has killed two birds with one stone; he has carried out a new test of Einstein's general theory of relativity and he has confirmed our suspicion that matter 2,000 times denser than platinum is not only possible, but is actually present in the universe."[7]

Adams's results were confirmed; much later, similar results were obtained regarding another white dwarf, Eridani, thus providing new measurements compatible with the Einstein effect, whose case had been strengthened with the passing of time. But the measurements concerning the Sun no longer supported the theory. In Bern, in 1955, it was believed that "the confirmation of the theory by the solar observations is not very convincing."[8] Thus, after forty years of difficult observations, the situation regarding the Einstein effect remained muddled. Its confirmation was tangled up in a complex network of different hypotheses, various effects, and poorly known magnitudes (hypotheses concerning solar mechanisms, the radius and mass of white dwarfs, and so forth); a network in which general relativity was not necessarily the weakest link. Everybody was anxiously waiting for new investigations to be carried out.

One can then easily understand the shock and the hope created by an *experiment* by Pound and Rebka in 1960, thanks to an effect recently discovered by a nuclear physicist that made it possible for them to perform the experimental verification of the line shift effect, which was soon improved to a precision of one percent! The experiment took place at Harvard at a height of only 22.5 meters. The good news was announced in the *American Journal of Physics* in an almost biblical tone: "These are exciting days: Einstein's theory of gravitation, his general theory of relativity of 1915, is moving from the realm of mathematics to that of physics. After 40 years of sparse meager astronomical checks, new terrestrial experiments are possible and are being planned."[9] Everybody welcomed these results, which allowed physicists to finally break away from four decades of gloom.

In 1971, new shift measurements regarding Sirius B were published. They largely invalidated the previous results because they assigned to the companion of Sirius a shift five times larger than the one obtained in the 1920s. But on the other hand they confirmed general relativity. Indeed, the ratio mass/radius that enters into the calculation of what is discreetly called "the theoretical prediction" had changed in the meantime and, following new measurements, in the same proportion as the spectral shift. Anyway, from then on general relativity had other trump cards in its hand.

A FOURTH TEST THAT COULD NOT BE FOUND

The story of the verification of the three classical tests of general relativity resembles the game of go. Just as in that game of strategy, it is a question of occupying space and holding one's position. Mercury's perihelion represented, in fact, a kind of minimal support on which the theory had stood since its creation. But the competition among other explanations was fierce and, even if they were far from being as convincing as general relativity, other theoretical constructions could explain that anomaly, as we have already seen.

After 1919 the situation changed: general relativity raised the stakes by extending its observational network and using other results to back its claims. Every rival theory (for, even if I don't mention it here, there was a certain competition going on in which several players took part) had to now also explain the deflection of light in a gravitational field and also the line shift. Relativity's position on the gameboard has improved. More or less seriously challenged, these two results went through some modifications until they seemed settled, only to be challenged again. But the perihelion stood fast, and no relativistic theory of gravitation could do better than Einstein's. In short, until the end of the 1950s the observational situation remained inconclusive but nevertheless favorable to Einstein's theory, which was undoubtedly the theory that accounted for the observations better than the others—even if not perfectly. And that's all that was required of it.

It was therefore through a complex process of mutual support that an empirical consensus regarding Einstein's theory was built and consolidated. The explanation of Mercury's perihelion formed of course, its foundation, due as much to the precision of the test as to the unexpected character of this success, and also to the total absence of arbitrary parameters. But starting in 1919, it was on the second test that the conviction rested, and when, later, the degree of precision seemed to be unsatisfactory, the perihelion once again became the main support for the theory.

All in all, the first and second tests complemented each other and together refuted Newton's theory and guaranteed general relativity's right of existence. In this process, other elements, theoretical (the increasingly accurate verification of special relativity that rendered a relativistic theory of gravitation even more necessary) as well as empirical (Eötvös's experiments regarding the principle of equivalence) played a role that, without always being perceived as crucial, was nevertheless important, and not only for Einstein.

As for the Einstein effect, it occupied a special place. More than any test, it was vital to the survival of general relativity even if it was hardly a *verification* of the theory. It was the least sensitive of the three classical tests, because many rival theories of gravitation also predicted such an effect. On the other hand, it was an extremely important test, for its failure would entail the refutation of a whole class of theories that included general relativity.

All through its history, general relativity has never been seriously proven wrong, and it succeeds decently and, in fact, better than any competing theory in accounting for the observational ground that it covers, a relatively small territory, to be sure, limited both by the available techniques and the scarcity of ideas for testing the theory. But within those bounds, general relativity's agreement with observations is quite satisfactory on the qualitative level, which does not mean that it covers that ground as well on the quantitative level.

The experts, however, unanimously lamented the particular difficulties for observing effects specific to the theory. Between the 1919 eclipse and the 1960 experiment—outside of the cosmological domain that we shall take up in chapter 15—the theory's observational range, still restricted to the three tests, remained disappointing. It was a deficiency, due above all to the amazingly small differences between its predictions and those of Newton's theory which, after more than two centuries of hegemony, left only a very tiny leeway for any other theory of gravitation to manifest itself empirically.

Numerous calculations were performed before the 1960 renewal in an effort to try to apply general relativity to other problems, such as the Moon's secular acceleration, the displacement of Mars' orbit, the atomic domain or the gravitational lens effect, the gravitational waves, the motion of the Earth's perihelion, and those involving the use of a gyroscope, not to mention problems in the cosmological domain (chapter 15). On the whole, these were thought experiments, given the difficulties in achieving the level of technical precision required for specific effects to be detectable. It was like a fourth test that could not be found!

To sum up, the situation was not exactly bright for its rare specialists, although not completely catastrophic for general relativity. The year 1955 was a decisive one in that respect; symbolically, because it is the year of Einstein's death, but also because it was at that time that a new generation emerged, one that painfully realized the state of their discipline, a state which their predecessors, whether they liked it or not, had to accept. The deep crisis experienced by general relativity left many scars in their minds, and everyone took part in analyzing the causes in order to try to move their discipline from under the shadow of Newton's theory. Experience—observation—was the central theme and a vital necessity, a topic that more than any other marked the beginning of the renewal. No book or journal failed to mention that the empirical predictions of the theory and its tests were still too meager nor to raise the question of the field in which the theory could develop.

Dicke, an experimental physicist coming from quantum physics, held an opinion that corresponded better than anyone else's to this desire for renewal. He had come over to the relativistic theories of gravitation with the firm intention to put Einstein's theory back on the right experimental track. Everywhere he went, he vigorously denounced "the paucity of experimental evidence,"[10] and, as a "distressing thing . . . the lack of contact of the theory with observational and experimental facts,"[11] and he announced his ambition to neutralize "the decided tendency in times past for General Relativity to develop into a formal science divorced from both observations and the rest of physics."[12] He undertook resolute action seeking to shift the center of gravity of research toward experience and the relativists' implicit idealism toward a phenomenological and positivist vision of the theories of gravitation.

CHAPTER 9

NOTES

1. Casimir, 1983, pp. 137–138.

2. Kuhn, 1962, p. 24.

3. Einstein to W. Julius, 24 August 1911, German Original in *CPAE*, vol. 5, p. 313; English translation, vol. 5, p. 199.

4. Einstein to Eddington, 15 December 1919, quoted in Earman and Glymor, 1980b, p. 199.

5. L. Silberstein, in Royal Society and the Royal Astronomical Society, 1919, p. 397.

6. Cf. box 5, chapter 7.

7. Eddington, 1926, p. 273.

8. R. Trumpler, in Mercier and Kervaire, 1956, p. 109.

9. Schild, 1960, p. 778.

10. Dicke, 1957, p. 363.

11. Dicke, 1962, p. 3.

12. Dicke, 1964, p. vii.

CHAPTER TEN

The Crossing of the Desert

In the first part of this book, I have presented general relativity's classical interpretation.[1] Classical in what sense? In the sense that it covers the first part of the life of the theory, from its birth in 1915 to the 1960s, and also in the theory's singular connections with that of Newton's. In the second part, beginning in the next chapter, I shall discuss its contemporary interpretation. But first, it will be useful to pause for a moment and examine the way in which the theory was experienced during its classical period, before the radical change of interpretation that took place during the 1960s. Experienced! The word is a bit pretentious, you may think, for a theory made of principles, equations, and observations. Not really. The primary element of a theory is the way it is conceived, understood, and interpreted—in a word, how it is *experienced*. A theory must be interpreted before it can be applied, and it cannot be defined only by stating its rules, rules that are infinitely more complex and fluid than they first appear. A theory only exists in the minds, opinions, and judgments of its experts, almost like a work of art. Isolated, forgotten, and with no theoretical or practical applications, the theory is like a lost book lying in a remote corner of a library, a book no one will need, read, or appreciate. In short, a theory lives. It exists to be understood by the minds of its experts at a given moment of its life. And, as new generations of experts come to understand it, the theory evolves, even if its equations do not change. The same is true of Van Gogh's paintings, whose significance changed in the course of time, or of Molière's plays, whose reality is that of their fervent followers and obstinate detractors. Is it possible that a piece of work, a book, or a painting only exists in relation to its public? (And its relation to its author, of course.) Take that relationship away and all that is left is the physical reality: paper, canvas, a bit of ink, oil and pigments in the back of an attic. This is the relativity of . . . relativity! Relativity would only exist relative to the relativists. This would be (almost) true if nature, if reality, did not have its say in words that the relativists are there to interpret.

Less than any other physical theory, relativity did not spring ready-made from Einstein-Zeus's head. Everybody had their say, defended their interpretation, and contributed to the evolution of the theory. For even if its fundamental equations remained unchanged (at least from 1917, when Einstein introduced his famous cosmological constant [see chapter 15]), the theory evolved. It did so under the pressure of the observational facts that it helped to identify, the pressure of the theoretical works and the applications to which it gave rise. It was a patent, striking, and dramatic evolution.

And I believe that the best way to understand the evolution of this interpretation, its necessity and its difficulties, is to listen to the clamor of the crowd. The crowd of relativists who applauded and worried, and that of many other theoreticians who frowned or grumbled. A contested and reviled theory? Was it possible? It certainly was. The reactions to Einstein's work ranged from ecstasy to severe condemnation, and all together they depicted the portrait of a lively theory. Isn't it worth a closer look? Sure it is. Let us listen to the crowd.

A "MAGICAL" THEORY

After Einstein, who, as we have seen, underlined the magic of his theory, the relativists were extremely sensitive to this strange architecture. General relativity was then widely referred to as a "model" of a physical theory, with emphasis on "its inner consistency" and the "logical simplicity" of its axioms. The theory was "the most perfect example of a field theory to date," an "ideal deductive and explicative physical theory," and "one of the greatest examples of the power of speculative thought." Attention was drawn to its "firm conviction power," and it was called "the greatest feat of human thinking," an admiration that the specialists shared with the epistemologists and philosophers seduced by the theory. Let us listen to the passionate analysis of a former collaborator of Einstein's: "What happened regarding the [general] theory of relativity is something strange that will perhaps endure. It is the fact that "metaphysical" thought, logical/constructive imagination, was able to penetrate the secrets of nature which we might not have been able to decipher in a purely empirical manner."[2]

In my opinion, this analysis touches on an essential point: in the construction of general relativity, Einstein did not start from experience. The anomaly of Mercury's perihelion, which was the only hitch in Newton's theory, was not at the origin of his analysis. Absolutely not. At the origin of his analysis was his conviction, soon to be shared by many of his colleagues, of the soundness of special relativity, and therefore that Newton's theory was incomplete and a *relativistic* theory of gravitation was needed. But, if this relativistic theory of gravitation had to be constructed, and there was practically no phenomenon to be explained, how to proceed? How was he to proceed except for those "metaphysical" views and that "logical/constructive imagination" that allowed him to sail through those principles whose touchstone is the principle of equivalence—to sail thanks to an "inspired imagination," and reach some unknown and improbable land.

There is a Faustian element in Einstein's work, especially in the way it is perceived—something surprising, moving, wonderful, or upsetting, depending on the point of view. Having started with so little, Einstein realized the essence of things, the structure of space-time, and the structure of the universe. All that from those simple little principles. He certainly got to the heart of the physical world, to the heart of things.

If, as a theoretical physicist, you consider the difficulty of finding your way among all those phenomena, facts, laws, and mathematical structures, there is reason to be stunned. More often than not, a new theory is based on a substantial body of physical facts, a solid theoretical field, and a great deal of observational data. When you consider that Newton, in constructing his theory of gravitation, had to account (among other things) for the planetary motions, or that Maxwell had to restructure and combine the whole of electricity and magnetism, general relativity, in comparison had practically nothing to explain. It was not going to account for anything (or almost anything), except those insignificant 43 seconds of arc per century of Mercury's perihelion, and it only predicted minimal or secondary effects. Despite all that, Einstein rebuilt Newton's theory and special relativity, which were working to everyone's satisfaction. How daring! How marvelous!

His theory worked well, very well, even if it only accounted for very few phenomena; even if it contributed—for the moment—very little to "real" physics, didn't it do a little better than Newton's theory and its rivals that already were legion? What else could we ask from it? But there were critics at work, with no weapons other than jealousy and bitterness, which generated disdain and contempt, but which did not prevent the relativists from praising their theory and sometimes playing into the hands of their detractors.

One of Einstein's closest collaborators believed that "*the most convincing* arguments in favor of the general theory of relativity, however, remained, so far, theoretical."[3] That was not quite true, for those theoretical arguments only made sense after observation had its say. But that person was neither the first nor the only one to give the positivists ammunition by suggesting that theoretical arguments may compensate for the empirical deficit. Here and there it was remarked that the reasons for preferring general relativity were not so much its agreement with observations as its almost complete lack of arbitrariness and the "confidence in its basic philosophy . . . which is intellectually appealing." Elegance, simplicity, lack of artificiality, intellectual appeal, and "basic philosophy" are the qualities attributed to Einstein's theory, qualities that contrast with the "unsatisfactory and capricious" character of the other possible explanations.

When Einstein, ordinarily so discreet in his scientific articles, let slip the qualifier *magical* in describing his theory, he expressed the formidable distance between the principles he relied on and their consequences, "the" perihelion being in this respect the most remarkable one. The—magical!—agreement between the calculated and the observed figures exceeded the degree of precision of the observations to the point of raising questions about the truth of the theory: "It is as if a wall which separated us from Truth has collapsed," exclaimed Weyl.[4] Presented with the success of an extremely improbable approach, astonishment and surprise became fascination. Had Einstein put his finger, if not on the "true" theory of gravitation, at least on some absolute principle?

It was hoped that this was the path leading to a "complete understanding of gravitation," which, however, could not be attained before the arrival of the field theory unifying gravitation and electromagnetism, on which Einstein and others were working already. Could this be the realization of their dreams, the unspeakable hope that was hatching in the minds of those who claimed to be firmly convinced by the magic of the theory?

In 1929, in an interview to the *Times*, Einstein explained the differences between general relativity and the other theories. He mentioned "the degree of formal speculation, the slender empirical basis, the boldness in theoretical construction, and finally the fundamental reliance on the uniformity of the secrets of natural law and their accessibility to the speculative intellect."[5] But, he added, "However that may be, in the end experience is the only competent judge."[6] Amid criticism, Einstein asserted himself with remarkable clarity; yes, his was a formal theory that, indeed, had little empirical basis—an audacious theory, but one that experience validated. And it was the exemplary success of the first two phases of his program (special and general relativity) that allowed him to risk the last stage of his unifying project, whose dynamics and deep significance he explained: his hope that the secrets of nature should be accessible to the "speculative intellect."

Elegance, harmony, internal beauty, and unmatched esthetics are expressions of the seduction the theory exerted on its specialists, who could not deny the pleasure they derived from a nicely constructed theory. Whether a motive for rejoicing or regretting, there is a reason why so many esthetical images were employed in describing general relativity. Unquestionably, the theory had much in common with a work of art—a rather abstract art, to be sure—because the representation of the phenomenon proposed by the theory was so remote from its original image and because of the radical, revolutionary character of the vision of the world it introduced. These were features of the theory which many scientists have emphasized, whether to praise it or to condemn it. Esthetic pleasure was a fundamental reason for working on Einstein's theory. It was, in a way, a necessary justification, given the scarcity of concrete results or physical effects that the theory offered to its followers (not to mention the lack of recognition and its repercussions on their professional career).

If the esthetic metaphor provided the most fervent relativists with one more reason to believe in the theory, it was also used by its critics to voice their misgivings and manifest their disapproval. "The theory of relativity of Einstein, quite apart from its validity, cannot but be regarded as a magnificent work of art,"[7] declared Ernest Rutherford. But, "apart from its validity," what was left of a physical theory if not its art and its style? The esthetic argument was thus turned against the relativists with a pejorative slant. For if relativity was *only* a work of art, then its specialists were *mere* artists that produced ideas—magnificent ones, to be sure, but of little use—a luxury.

Rutherford feared that Einstein's theory would drive many scientists toward metaphysical conceptions, diverting them from their calling and keeping them away from *real science*. And he added: "We already have plenty of that type in this country and we do not want to have many more if Science is to go ahead."[8] And when the German physicist Wilhelm Wien supposedly commented to Rutherford that "no Anglo-Saxon could understand relativity," Rutherford agreed readily: "No, they have too much sense."[9]

This aspect of the theory may be added to its mathematical complexity and to the chasm between the theory's basic hypotheses, on the one hand, and the directly observed facts, on the other. Plenty of authors would call attention to the gap—the opposition between the logical simplicity of its foundations and the complexity of its technical implementation—and also to the effort required from everyone to question the classical and familiar structure of space and time. All of this was held against the relativists, who were accused of being first and foremost mathematicians, an unforgivable sin in the eyes of a positivist. Numerous physicists would feel, or would want to be, excluded, and some of them would go as far as to quietly admit their inability to understand the theory.

"Since the mathematicians have invaded the theory of relativity, I do not understand it myself any more."[10] So Einstein himself put it after Minkowski had proposed his four-dimensional formulation of special relativity, which would soon become the standard one. There is no doubt that Einstein's former mistrust of mathematics was found in some of his colleagues; plenty of physicists were not convinced of the advantages of mathematizing physics. And the announcement of the verification of Einstein's predictions created a real panic among the physicists who feared they would have to study the theory of tensors. One of relativity's strongest opponents, Sir Oliver Lodge, saw "terrible times" coming for the physicists.[11]

"This involves much very abstruse mathematics, and there is much of it I do not profess to understand," declared J. J. Thomson,[12] for whom "our ultimate aim must be to describe the sensible in terms of the sensible."[13] But more than an admission, this was a means of denouncing the incomprehensible nature of general relativity: a poisonous and scathing remark that sought to shut down the theory and its ghetto of specialists with their abstruse language. The difficulty issue was used in the same way as the esthetic argument. And as esthetics became abstruseness, the difficulty became incomprehensibility: a double rejection. The "incomprehensibility" of Einstein's theory was, in fact, a blow below the belt coming from those who, with no particular interest at stake and not having had the time or the desire to study it, worried at their total lack of understanding. In order to justify themselves, they denounced it: "It is an incomprehensible theory!" Translation: this theory is of no use to me, and I don't need or wish to understand it.

"Apart from its validity," to employ Rutherford's expression, general relativity was an abstruse and metaphysical work, of interest only to esthetes; a confidential and incomprehensible theory, and therefore useless to the ordinary physicist who favored a physics close to experience; a piece of figurative art. The firm conviction of this commonsense man was forcefully expressed by a French physicist:

> The reason for this glory, which I'm afraid will not last, is that Einstein's theory does not fit the pattern of a physical theory: it is a *metaphysical hypothesis* which, in addition, is incomprehensible, two reasons that explain its success. . . . The wild applause of a bunch of incompetents adds nothing to the admissibility of a hypothesis. . . . We couldn't care less if mathematicians and astronomers, considering themselves superior, call us ignoramuses and uncultivated, and end up by insinuating that we are senile, ripe for the wheelchair and the bib. All those kind words leave us cold, *because at the end of the day we, the experimental physicists, will have the last word*: we accept the theories that we find convenient; we reject those which we cannot understand and that are for this reason useless.[14]

This is passion speaking: the same passion, the same fanaticism present in the petit bourgeoisie and which made them hate those cubist, Dadaist, and nonfigurative paintings of the turn of the twentieth century. The same ones who boasted about not understanding the art nouveau that the snobs applauded without understanding anything. Wasn't incomprehensibility the hallmark of that art? It was also a label that general relativity had to endure as a means of pushing the relativists into a ghetto. In short, the esthetics metaphor seduced everyone, opponent or follower, and fueled many a speech; it was at the heart of the ideology that then pervaded the theory.

But apart from the aggressiveness of some theoreticians, one can also explain this rejection by the desire to end gravitation's isolation, not just with respect to quantum theory, but also in regard to the whole of physics. The geometric *pedestal* on which Einstein's theory so complacently stood was challenged as well as coveted; it was a pedestal that represented a certain ideological dominance of this revolutionary theory but that was also a symbol of its isolation.

EINSTEIN'S THEORY AND QUANTUM MECHANICS

After the 1920s, Einstein spent little time on general relativity. He was "not content with his creation"[15] and in particular with the description of the field sources, "a makeshift," he would write in his autobiographical notes,[16] to the point that he wondered, "[W]hy should anybody be interested in getting exact solutions of such an ephemeral set of equations?"[17] For him, general relativity represented only "a way station" on a path which, starting from special relativity, would lead to a unified theory of gravitational and electromagnetic interactions. "It seemed quite possible to him that a suitable unified theory might also explain the various quantum effects then being so rapidly discovered," wrote J. Stachel.[18]

In a 1921 special issue of *Nature* devoted to the theory of relativity, Einstein distinguished "some of the important questions which are awaiting solution. . . . Are electrical and gravitational fields really so different in character that there is no formal unit to which they can be reduced? Do gravitational fields play a part in the constitution of matter, and is the continuum within the atomic nucleus to be regarded as appreciably non-Euclidean?"[19] For Einstein, these were not so much questions as hypotheses of a single program that he would relentlessly pursue.

Actually, it was not only his project but also the shared endeavor—after the initial ideas of Arthur Eddington and Hermann Weyl—of a group of theoreticians close to him. The program entailed not only a perspective based on extremely restrictive principles, which made it highly demanding, but also a repeatedly thwarted expectation that was far from being unanimously accepted.

At the beginning of the 1930s, Einstein published a new "unified physical field theory" which, in his own words, aimed "at the renewal of the general theory of relativity" but which for the time being was only "a mathematical construction barely connected to physical reality by some loose links"[20]—a theory that prompted Pauli to wonder "what had become of the perihelion of Mercury, the bending of light, and the conservation laws of energy-momentum"[21] without eliciting a satisfactory reply. It was then that Einstein wrote in a philosophical journal: "I do not consider the main significance of the general theory of relativity to be the prediction of some tiny observable effects, but rather the simplicity of its foundations and its consistency."[22]

And it is precisely what he expressed even more clearly in the preface to Bergmann's book that he wrote in 1921: "It is true that the theory of relativity, particularly the general theory, has played a rather modest role in the correlation of empirical facts so far, and it has contributed little to atomic physics and our understanding of quantum phenomena. It is quite possible, however, that some of the results of the general theory of relativity, such as the general covariance of the laws of nature and their nonlinearity, may help to overcome the difficulties encountered at present in the theory of atomic and nuclear processes."[23]

Wasn't it characteristic that Einstein should consider general covariance to be a "result" rather than a tool of the theory? This was because for him experience was not an end in itself but a means that, through general relativity, allowed him to validate his principles (covariance, in this case)—principles that represented a step toward a unified theory, his real goal.

But, beyond the speculative character of these perspectives, beyond Einstein's deepest motives, the risky gamble on unified theories also represented a way out for general relativity—its principles and methods—and for certain specialists who would then be able to find research topics. It was a very risky gamble indeed, whose unpredictable payoff would be a fantastic one, for it involved, at least at the beginning of the 1920s, nothing less than constructing a single physical theory for the two interactions known at the time—gravitation and electromagnetism.

To those who may have questioned the interest, or even the possibility, of a geometrization of (all of) theoretical physics, the "alternative" theories of gravitation (see below) provided another way out, a priori less unpredictable than the unification theories. Those two approaches were irreconcilable, two antipodal projects for the same diagnosis: the need for a connection if not unification between physics' two fundamental interactions; two perspectives, for which general relativity was the reference. However, these multiple attempts at the unification of the gravitational and electromagnetic interactions would prove problematic,[24] and the particularly formal works would accentuate the isolation of the specialists who carried them out, often diverting them from the study of general relativity.[25] But not everybody agreed with this analysis. Cornelius Lanczos, who had himself participated in this kind of work, was not as harshly critical of the program, as he explained in an article retracing Einstein's unifying aims:

> But if we may regret that he ostracized himself to strange countries to which nobody was willing to follow him, we have to admire the intellectual honesty with which he pursued his aims. Einstein-Faust gazed at the sign of the Macrocosmos for ten years, trying magic formula after magic formula, to conjure up the mighty spirit of creation. And then the magic formula was found and the mighty spirit did appear in all its splendor. Can we blame him if he could not settle down again to the gray daily chores of life? Can we expect that a man who has seen face to face the cosmic apparition, should help little people in building little homes?[26]

The "way station" that general relativity constituted for Einstein can thus be explained by the wider horizons he saw for his research; Einstein's isolation reflected the perception that many physicists had of general relativity and the relativists.

We must of course recall the complex and difficult relations that Einstein maintained with quantum physics and its specialists—his dissociation from the supporters of the Copenhagen interpretation that soon prevailed. His position was symbolic of a very idealistic conception of theoretical physics, an issue that marked, through his philosophy, Einstein's image in the scientific circles as well as the image of his general relativity.

Einstein's personality, his world vision, the image of physics he projected, his demands and his refusals, and his loneliness that were typical of the second part of his career—all those things could not fail to have had an influence on those who, from the mid-1920s to World War II, decided to work in general relativity. The few theoreticians who made that choice generally produced relatively formal research work, either because they more or less shared Einstein's global vision or because they were led in that direction by the theory and forced by the smallness of its empirical domain.

A demanding theoretical emphasis, a commitment to unification, a marked epistemological interest, scientific production of a relatively formal character, the refusal of a phenomenological conception of the theoretical construction, a heavy reliance on mathematical structure to counterbalance the weakness of the empirical connections, and the rudimentary nature of the research structures—such are the main characteristics of the relativistic production, attributes that are not exactly present in the quantum field.

This was a real divorce, not only between the specialists of gravitation and those of the theories of the infinitely small, but also and especially between those two grand branches of theoretical physics. A divorce that, well beyond ideology—idealist among the relativists, positivist among the atomists—marked theoretical physics for a long time, for it was based on an extremely serious fact, a kind of fracture that divided the field: the fact that the space on which the theoreticians of the infinitely small work, and in particular those in quantum theory, is essentially different from the space required in gravitation. No one had a problem with accepting, more or less promptly, special relativity, and so everybody worked in the Minkowski space-time that this acceptance implied. But the specialists of the infinitely small felt helpless before the curved space-time on which general relativity is based. They had no use for a Riemann space in which their tools were, for all practical purposes, ineffective.

The specificity of the relativistic domain can only be fully appreciated when compared with the conditions then prevailing in quantum physics: a particularly lively theoretical domain, an implicit philosophy considerably different from the one favored by the relativists; and, above all, the participation in a booming experimental field: particle physics. Quantum mechanics was going through an unparalleled development—to which, let us recall, Einstein had made brilliant contributions—that showed no sign of running out of steam.

Just consider: Planck's quantum hypothesis (beginning of the twentieth century), Einstein's photoelectric effect (1905), Bohr's atom (1913), Stern's and Gerlach's experiment (1921), de Broglie's wave-particle duality (1923), Heisenberg's matrix mechanics (1925), and Schrödinger's equation (1926). But there is more: Dirac's quantum electrodynamics (1927), the first accelerators (beginning of 1930s), the discovery of the neutron (1932). Is it necessary to go on? It was an impossible competition for general relativity, whose potential theoreticians could come only from theoretical physics. For any young, or not-so-young, theoretician, the choice of a dynamic field was crucial—and obvious.

The extraordinarily different characteristics of these two theoretical fields are made even clearer if we consider work environments and research facilities. On the one hand, we find theoreticians often attached to leading laboratories working together on theories in constant evolution and in contact with a large number of experimental scientists who use increasingly powerful machines. On the other hand, there were university professors working, more often than not, alone on some mathematical aspect of Einstein's equations.

As for the number of publications, the picture was no different. If, in this respect, the importance of general relativity at the beginning of the 1920s was approximately equivalent to that of quantum mechanics, it was reduced before the war to a small fraction that is difficult to evaluate because of the multiple interconnections of quantum physics at both the theoretical and experimental levels and in their applications. Around the time of Einstein's death, we find some thirty references per year to general relativity in *Physics Abstracts*, out of a total of almost ten thousand—that says it all.

Einstein was well aware of all this. In 1949, he wrote to Max Born, his friend and a leading specialist in quantum mechanics: "But our hobbyhorses followed different paths and are now hopelessly far apart. The fact remains that yours enjoys, naturally, a much greater popularity given its considerable practical achievements, while mine has a quixotic look and even I do not give him absolute credit."[27] It was a contrast that was probably still present in Born's mind when he recalled, at the Bern conference six years later and only a few months after Einstein's death, his memories and his choices (see chapter 11). As a young physicist, Born decided "never to attempt any work in this field"[28] (i.e., general relativity), a decision also made by many of his colleagues, even if some of them—including Born—would make, sooner or later in their career, a contribution to Einstein's theory,[29] which a few, such as Pauli, would later regret. What Born and many others like him did, whether consciously or not, was to make an economic choice, a choice favoring productivity,[30] in the sense that they preferred to work in a field where the profit/investment ratio appeared most favorable on many levels: on the intellectual level—rapid scientific development and heated controversies; on the sociological and psychological level—abundant exchanges of ideas and important benefits, in a larger sense; and on the institutional level—more generous budgets and better job opportunities. In addition, Born made a philosophical choice: to the "realistic metaphysics," as Lanczos called it, on which Einstein's theory was based, he preferred the positivism to which the quantum field is infinitely closer. However, it was perhaps even more a crucial choice because compared to the asceticism of relativity, which was then extremely introverted, the quantum theories had links with the whole of physics and the world.

Those choices, those reactions, left very little room for the epistemological influence that Einstein's theories had on establishing quantum theory. The notion of fertility, of productiveness, is relevant not only on the experimental level or in mathematical physics, but also, and perhaps even more so, in the domain of theoretical constructions, a particularly important point that Wolfgang Pauli raised in the preface, written in 1956, to the translation-reissue of his 1921 book,[31] and therefore at the time of the Bern conference that he chaired. Pauli insisted on the fundamental role played by Einstein's theories as examples, and how they had represented "the first step beyond a naïve visualization." According to Pauli, Einstein showed the way that led to abandoning, in the theoretical framework of quantum theory, both "the concept of classical field and that of orbit of a particle."

> Without this general critical attitude, which abandoned naïve visualizations in favour of a conceptual analysis of the correspondence between observational data and the mathematical quantities in a theoretical formalism, the establishment of the modern form of quantum theory would not have been possible.[32]

And, in closing his preface, Pauli wrote: "I consider the theory of relativity to be an example showing how a fundamental scientific discovery, sometimes even against the resistance of its creator, gives birth to further fruitful developments, following its own autonomous course."[33]

Pauli insisted, quite rightly, on Einstein's "critical attitude," an attitude that influenced the establishment of quantum theory. In fact, the interest and the influence of general relativity went far beyond its empirical results and its equations. It was an example that made its mark: no one had until then dared to create a working space so innovative and so revolutionary. In this respect, it is clear that Einstein brought a freedom of thought that completely revolutionized the methods of theoretical physics.

THE NEWTONIAN DOMINATION

The closeness of Newton's and Einstein's theories of gravitation, in the sense that the Newtonian theory sufficed to account for the vast majority of gravitational phenomena, was then one of the most worrisome and disappointing questions. During the course of more than fifty years, the relativists were only able to snatch a few decimal places from Newton's theory and credit them to their own, a few decimals which, as we have seen, would be extremely difficult to consolidate with new evidence. Even if they obtained two *new* effects, it was little, very little, for such a weighty theory demanding so much effort!

Worse still, the relativists got trapped by Newton's theory: by the manner of looking at things and the calculation methods that it generated, and by their being too used to thinking in that context. Above all, of course, Newton's theory was enough to explain all that needed to be accounted for at the time: basically, phenomena in the Sun's own "backyard." Without really being aware of it, the relativists implicitly accepted using Newton's theory as a starting point for deciphering Einstein's. More precisely, their results often referred to "the post-Newtonian approximation," a technical term that means exactly what it says: that they started from Newton's theory. To sum up, the contribution of general relativity being (at that time) almost negligible, it was expected a priori to be small by implicitly acknowledging that nothing essentially changed if one stuck to the Newtonian description. This position condemned the relativists to never making any significant discoveries. They worked with mathematical techniques based on the Newtonian theory, and this approach imposed its own logic. There was nothing wrong with it as long as the prediction of some small effect was concerned, as was then the case. But this approach allowed them to detect only very weak effects and was certainly not designed to discover typically relativistic ones, which are of an entirely different order and obey a different logic.

We should not be too quick to blame our experts, though, not only because the appropriate techniques were not available to them but especially because they did not clearly see what Einstein's theory could do for them nor, above all, to what kind of phenomena it could apply. At the time, no (stellar) object massive enough for general relativity to play a noticeable role was known. As we already saw (box 5 in chapter 7), in most cases a factor of order $2MG/ac^2$ is at work, which is too small for any object in the solar system. This is of order 10^{-6} for the Sun (where M is the Sun's mass; *a*, its radius; and G, the gravitational constant); its square means that the effects would be of order 10^{-12}, hence negligible (with respect to 1) and impossible to observe. So, why bother? The fact remains that after order "zero," if I may call it that, which accounts for Newtonian results, the next order, or first order of approximation, is both sufficient and essential, because it is, on the Sun, for example, the order of magnitude of the Einstein effect.

Even worse, the domination of Newton's theory was felt on another level: that of interpretation. We have seen that Einstein's theory allowed only the correct definition—that is, in an intrinsic manner—of very few physical magnitudes, and it was not easy to properly define magnitudes that were truly relativistic. Our experts therefore *imported* from Newton's theory observable magnitudes such as distance, mass, energy, and so forth. To do so meant to generalize, that is, to render relativistic, those Newtonian magnitudes, but this approach often led to disastrous results, because it was not so simple to associate to each Newtonian magnitude a corresponding "Einstenian" one. Why this was so is difficult to say in general, but it was probably due to some fundamental reasons. Today, experts generally agree that a definition of "distance" is neither necessary nor desirable, simply because it would require a notion of rigidity, which is itself based on the discarded idea of simultaneity. In general relativity there is no distance in the classical sense; a notion specialists can do without thanks, precisely, to the concept of proper time. Despite all this, relativists often, too often and for much too long, worked with and thought in terms of a *neo-Newtonian* distance, resulting in a great deal of confusion, for no one knew exactly what it meant, except that it could not be an *intrinsic* definition, that is, independent of the way it was measured.

In short, plenty of experts needed many years to learn how to do without distance, but the case of distance was not an isolated one. Energy, for instance, was another concept that could not be easily defined in an appropriate manner. Likewise, the definition of an even more fundamental concept, that of space-time, was not given a priori and it required rethinking the whole question, realizing that space-time really had to be *invented*. This invention could not be achieved without new tools that were far from obvious and ready to use. It was necessary to invent answers to questions that arose in the theoretical domain as well as in the models and applications of the theory. Work in this respect was carried out on two fronts and allowed the relativists to perceive the real questions. The first front was the cosmological domain, which raised the question of the true structure of space-time. And later, much later, the search for a solution to the (apparently easy) question of the structure of space-time created by the presence of a spherical mass, a solution in which the invention of the concept of *black hole* would play a fundamental role.

In a nutshell, everything had to be rethought and rebuilt far from the Newtonian ideology. This would take time, and a lot of it—the time for the relativists to convince themselves and change their mentality. In the meantime, they worked with the neo-Newtonian interpretation of general relativity, which was sufficient to handle the phenomena in the Sun's vicinity. But those specialists locked themselves in an outmoded version of Einstein's theory, a fact that was reflected almost daily in the literature from 1920 to 1960. This was perhaps inevitable, but it resulted in the implicit acceptance of a neo-Newtonian interpretation of the concepts. In resigning themselves to this state of affairs, the relativists simultaneously renounced following through on general relativity; it was only when the theory encountered limitations and singularities on the cosmological, astrophysical, and quantum levels, that it gained a fresh impetus. We will come back to these issues in the coming chapters.

Despite the above considerations, there were some relativists who did devote part of their works to the "pure" theory of relativity, being perfectly aware of the need for a specifically relativistic approach. "The value of this theory clearly lies in the new insights it provides and not in each of the corrections it makes to Newton's equations,"[34] wrote one expert. It must be emphasized how difficult this approach was and still is; and it certainly did not help that no specifically relativistic experiment was possible until the late 1960s, the type of experiment that would have enabled the relativists to find the right direction. Necessity knows no law; the practice of the theory for a long time remained mostly neo-Newtonian.

In 1962, at the end of a conference "The State of Relativity," two leading specialists of the theory, H. Bondi and J. L. Synge, resorted to a bit of humor in treating the subject. Let us listen in on them:

> *H. Bondi*: I should like to take up Professor Synge on one point; I am not entirely in sympathy with his hankering after reality. It seems to me that a great deal of the interest of general relativity lies in asking what the theory would say in conditions which admittedly do not occur in those parts of the universe about which we know much. In conditions which are common, we know that Newtonian theory is a good approximation, except for some minor points of which we had a very clear description this morning. Given that this is so, I feel that Newtonian theory largely satisfies my own hankering after reality.
> *J. L. Synge*: Is Professor Bondi satisfied with Newtonian theory—philosophically—with its absolute time?
> *H. B.*: Oh no! I was not looking at Newtonian theory as a satisfying theory. I was looking at Newtonian theory as a particular and well-worked-through approximation to the relativistic equations, and I felt that this particular method of approximation—admittedly invented 250 years before the theory—nevertheless, is highly successful in practically all cases appertaining to reality.[35]

The Newtonian theory: an outdated vision that still worked perfectly well and with which one should be satisfied. Were the best experts throwing the towel?

Over the years, many relativists would worry about the technical difficulty and the complexity of general relativity. Leopold Infeld, a close collaborator of Einstein's, observed "how difficult the mathematical deductions are, how complicated are the equations of general relativity theory and how deeply they can hide their secrets."[36] It was an objection that the new generation would not share and which John Wheeler, one of the artisans of the theory's renewal, would deride in 1957, when he wrote, in an ironical tone: "An objection one hears raised against the general theory of relativity is that the equations are non-linear, and hence too difficult to correspond to reality. I would like to apply that argument to hydrodynamics: rivers cannot flow in North America because the hydrodynamical equations are non-linear and hence much too difficult to correspond to nature."[37]

Wheeler would be the leader of a new dynamics opening up some theoretical and astrophysical possibilities in opposition to the neo-Newtonian interpretation of the theory, as the following assertive passage from a technical book published in 1965 illustrates: "Except for the prediction of Einstein's theory about the expansion and recontraction of the universe, all the other applications of general relativity (precession of perihelion, redshift, bending of light, gravitational radiation) as normally envisaged have to do with small departures from flatness. Not so here. Collapse produces geometries almost as far as can be from flatness. . . . If one intends to abandon relativity, here is the place to do so. Otherwise he is on the way into a new world of physics, both classical and quantum. Here we go!"[38]

This is precisely what we will do in the next chapters.

IN SEARCH OF AN ALTERNATIVE THEORY

While relativists waited for the renewal, a search was on, in a more classical spirit, for a theory that would be easier to handle. Dozens of articles were devoted to the numerous "alternative" theories to general relativity—more or less serious alternatives, it must be said. They were relativistic theories of gravitation, a sign that Einstein's initial project was accepted, if not the solution he proposed, because an "alternative" theory can only be born out of a genuine dissatisfaction with Einstein's, and not without ulterior motives. All these theories were built around special relativity, but their analysis was ultimately based on empirical predictions: the three classical tests (perihelion, deflection of light, and spectral shift) to which certain theoretical predictions not yet experimentally verified were added. But why on earth look elsewhere for something that was already found? Everyone agreed, even if reluctantly, that Einstein's theory had all the qualities of an excellent physical theory, perhaps too many qualities. The proof of this was its role as *the* reference, against which the other theories, called "alternative" for a reason, were measured. But what, then, were the charges against Einstein's theory of gravitation?

An analysis of the objections to general relativity made by those who, between the two wars, attempted to build another relativistic theory of gravitation revealed two quite distinct issues: on the one hand, a criticism of an epistemological, if not philosophical, nature questioning the principles of the theory; on the other, and this is the essential problem, the Riemannian and nonlinear character of general relativity, which was responsible for the difficulties in the technical implementation of the theory. But no one questioned special relativity or the ability of general relativity to answer the questions that experience had asked so far.

"It is inherent in my theory to maintain the old division between physics and geometry. Physics is the science of the contingent relations of nature and geometry expresses its uniform relatedness," wrote Whitehead in his book on a competing theory.[39] And, if he declared himself the heir of special relativity, he rejected the Riemannian framework, a position he shared with many of his colleagues. To the liberty Einstein took in restructuring his theories according to the principles he believed in, they opposed the liberty of remaining faithful to the traditional framework.

The principle of covariance was the favorite target of one of those authors who fiercely attacked Einstein's theory with a contemptuous tone through which (we are in 1940) a certain anti-Semitism shows: "The mysticism which Einstein in his 'special' relativity cleared out by the front door when he insisted on observationally determined numbers alone being used to fix events, returned by the window when he introduced general coordinates in 'space-time.' 'General' relativity involves a form of atavism; we shall obtain in due course a generalization of 'special' relativity in its natural line of development."[40] And, elsewhere, in a very colorful way: "General relativity is like a garden where flowers and weeds grow together. The useless weeds are cut with the desired flowers and separated later. . . . In our garden we try to cultivate only flowers."[41]

These views expressed the essential fracture, the personal violence, that the loss of such a fundamental reference frame as Newton's absolute space represented for certain physicists. And it was above all the Riemannian structure of space-time that was in question: "Also Einstein's law of gravitation, amazingly successful and satisfying as it was, was by no means an inevitable consequence of the conceptual basis given by describing the phenomena by means of a Riemannian metric. I have never been convinced of its necessity."[42]

The problem was not with the results of general relativity, nor with its predictive value or some particular weak aspect of it; it was the fundamental structure of the theoretical apparatus itself that was rejected, probably due to the huge gap between the structure at the origin of the theory and the empirical facts, between the imagined space and the space really experienced.

I would be tempted to see in the aggressiveness displayed by so many physicists, a response, largely unconscious, to what they perceived as a painful loss of their traditional bearings, that is, the frame of reference that the Newtonian absolute space constituted. The very structure of their thinking processes was affected in a profound and intimate way, and they reacted to this loss of reference system, which had been provoked by the questioning of their space, their position, their familiar markers in the deepest sense.

Relativity deals with some tricky concepts that the ordinary physicist finds difficult to understand. These concepts are time and, probably even more so, space, which we directly experience and believe to have an unclouded understanding of. Our daily experience with time and space structures our entire life, our comings and goings, our steps, our culture, and probably also our brain processes. We have grown so accustomed to our familiar perception of time and space that we find it very difficult to call it into question.

This was also the case with Newton's mechanics and theory of gravitation. Anyone who has plowed through a geometric proof in his *Principia* can understand the extent to which it was then difficult to figure out those proofs: results and theorems are usually expressed as ratios between areas of geometric figures, which does not facilitate the reader's comprehension. It was a question of culture, that of the use of geometry for calculating and proving, shared by a small circle of specialists. Algebra, which nowadays simplifies the solution of those questions, would conceal the problem, for we would no longer always clearly understand, or *see*, each step of an algebraic proof. The latter works a little like a computer: we input data and obtain an output not easily analyzable in terms of the input.

I cannot resist the temptation to quote here a *natural philosopher*, Jean-Paul Grandjean de Fouchy, secretary to the French Royal Academy of Sciences who, in the middle of the eighteenth century, rejoiced that algebra only required a simple calculation but that then a "significant drawback" remained: "For if the algebraic calculus is very appropriate," he wrote, "for facilitating the solution of problems," it is not so "in enlightening the mind on the processes at work" and, he added, "one is, so to speak, transported as if in a swarm to the solution, without having seen anything along the way."[43] *Seeing what is along the way*: that is called understanding. Understanding, or seeing, is what Newton's geometric proofs allow us to do, though not without difficulty, by following them step by step, but this understanding is what algebra partially hides.

As for relativity, aren't we confronted with an (almost) total loss of representation? To the loss of certain concepts that we had taken for granted, first of all, that of distance, (a visual concept) and, of course, that of space. Space-time, whose immediate representation is almost completely beyond us, because the non-Euclidean character of space prevents any *spatial* representation, essentially, because the view I have, for example, of this sheet of paper, presumes a Euclidean structure that is irreducible to that of space-time. We are thus condemned to forgo immediate representations, to no longer simply *seeing*. In this respect, the situation is similar to that of cubist paintings breaking away from figurative art. We must accept the loss because we have no choice and because we have other tools to go forward "in a swarm," without seeing anything along the way; tensors are a good example. Even if it means to later rebuild a representation of that other world, a weltanschauung.

CHAPTER 10

EINSTEIN COUNTERATTACKS

Confronted with those concerns, with those attacks, general relativity resisted them and did so quite well. Einstein did not content himself with defending the principles of the theory and emphasizing its merits; early on, he asked himself the key question of the validity of a physical theory in relation to competing theories. He was well aware of his theory's struggle for survival and started out to set the stakes on the epistemological level. He knew the temporary, mortal character of physical theories. Hadn't he just struck a terrible blow to the greatest, the most famous physical theory—that of Newton? It was a wound of immense symbolic meaning.

Once again, he chose the critical approach: it was not a matter of knowing whether a theory was true or not, because the question of truth is meaningless. A theory is not eternal; theories are mortal. We must give up hope of ever reaching the (scientific) truth. For the moment, the only relevant question is to make sure that one's theory has not (yet) been contradicted by experience. These are not simple ideas and Einstein was going to reflect on the question of the discovery, the invention, and the verification of physical theories. This epistemological analysis would be very helpful, for it allowed him to better see the real position of his theory with respect to Newton's, far from certain gesticulating colleagues. In short, amid all that, what was the state of gravitation?

In December 1919, immediately after the results of the eclipse were reported by the English expedition, Einstein wrote in a popular German journal: "But the *truth* of a theory will never be proved. Because one can never know if, in the future, some experience will become known that would contradict its conclusions."[44] "Never the truth": a loss we have to accept.

On the chessboard of existing theories, either proposed or likely to be proposed, Einstein helped to clarify the rules of the game and formulated new ones—a game whose real arbiter was experience; a game with no end because there was no final answer to the question. This point of view restricted de facto what a theory had to be but also provided arguments to defend general relativity. Some years later, he returned to this issue, in a clearer and simpler way: "[Experience] never says 'yes' to a theory, but, on the contrary, at best it says 'maybe,' and in most cases, simply 'no.' "[45]

A theory cannot be *true*, it can only be *right*. And of course, this new formulation of the criterion for the validity of a theory, this novel way to determine the *interest* of a theory condemned that of Newton, because experience had twice said *no*: in connection with the perihelion and with the deflection of light rays. But this did not necessarily mean that general relativity was *true*, for experience had only said "maybe." The new theory remains, therefore, today the least flawed theory of gravitation possible; a harsh way to put it, but can we ask for more? All in all, theoreticians do not have the right to reject a consistent theory that does better than the others and to which experience has not said "no." But they cannot tell what the future holds, the observational as well as the theoretical future. All they can do is to find a new theory that works as well if not better, by proposing and performing observational tests that relativity, for its part, cannot overcome.

These are familiar ideas today; the same ones in fact that, after Einstein, Karl Popper formulated under the name of the refutability principle.[46] Popper acknowledged, on several occasions,[47] the influence that Einstein and the example of general relativity had on his own views. Actually, even if they were far from being clear in everybody's mind, these ideas were "in the air" before the 1920s.

Einstein thus skillfully constructed, and at the right time, a defensive wall that would allow his theory to resist the competition from "alternative" theories, the scarcity of its applications, and the weakness of its experimental verifications, and that at the same time would open some opportunities for theoretical research. As far as Einstein was concerned, it was not as much a matter of *discovering* the laws of the universe as of *inventing* them.

At the time, it was generally believed that it was sufficient to observe nature with enough concentration to be able to advance toward truth, thanks to the so-called induction principle. Starting from the observation of natural phenomena, one would thus discover their regularity (such as, to give a simple example, the laws of the oscillation of a pendulum), and from those observations, one would be able to uncover the general laws of that particular field. In a way, the scientist only discovered the laws of nature; he or she was a revealer, not an inventor.

Einstein disagreed with this point of view; for him, there was no sure path leading to the basic laws, to the theories. A theoretician was *free* to invent the theory on which he would try to base the field in question; but later on, he would absolutely need to link his theory to the empirical domain, and the theory would have to account for the observations better than the one it was expected to replace. Curiously, "freedom" is one of the most common words in Einstein's epistemological writings. He strongly defended this freedom to invent—this possibility to use intuition—throughout his entire life. But there was a limit to the invention of the framework, the concepts, the principles, and the equations of his science; an absolute limit to which everything must be sacrificed and which was "the supreme arbiter"[48]: experience. At the end of his life, he drew an extremely enlightening diagram of the theoretical physicist's method for the benefit of Maurice Solovine, his old friend from the Olympia Academy.[49]

At the foot of this triangular diagram there is (E) reality, nature, the experienced facts and events that form the basis of the our intuition and starting from which, we must construct a possible theory. But the required theoretical elements—working framework, methods, concepts, structures—we can pick from here, there, and elsewhere, and not necessarily from the annals of science; we can invent them, for we are not, at this stage, bound by any accepted ideas or obvious facts. We then project and order those elements in a mathematical theory structured by axioms (A). This theory reflects reality in a mathematical context in order to account for, explain, and predict phenomena. From (A), we therefore descend toward reality, toward experience, by developing all possible logical consequences of the theory (A); this is the third side of the triangle. We deduce propositions, in fact, the predictions of the theory (S), (S′), (S″). Finally,

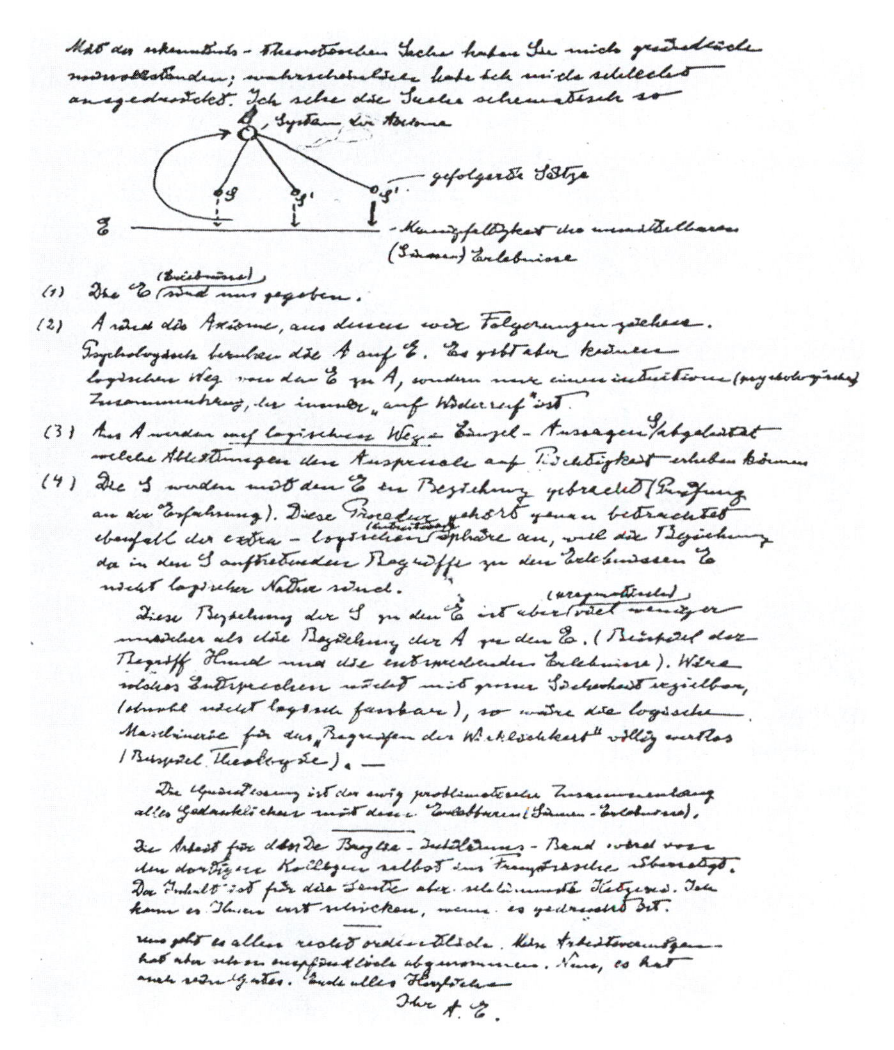

Mit der erkenntnis-theoretischen Sache haben Sie mich gründlich missverstanden; wahrscheinlich habe ich mich schlecht ausgedrückt. Ich sehe die Sache schematisch so

(1) Die E (Erlebnisse) sind uns gegeben.

(2) A sind die Axiome, aus denen wir Folgerungen ziehen. Psychologisch beruhen die A auf E. Es gibt aber keinen logischen Weg von den E zu A, sondern nur einen intuitiven (psychologischen) Zusammenhang, der immer „auf Widerruf" ist.

(3) Aus A werden auf logischem Wege Einzel-Aussagen S abgeleitet, welche Ableitungen den Anspruch auf Richtigkeit erheben können.

(4) Die S werden mit den E in Beziehung gebracht (Prüfung an der Erfahrung). Diese Prozedur gehört genau betrachtet ebenfalls der extra-logischen (intuitiven) Sphäre an, weil die Beziehung der in den S auftretenden Begriffe zu den Erlebnissen E nicht logischer Natur sind.

Diese Beziehung der S zu den E ist aber (pragmatisch) viel weniger unsicher als die Beziehung der A zu den E. (Beispiel der Begriff Hund und die entsprechenden Erlebnisse). Wäre solches Entsprechen nicht mit grosser Sicherheit erzielbar, (obwohl nicht logisch fassbar), so wäre die logische Maschinerie für das „Begreifen der Wirklichkeit" völlig wertlos (Beispiel Theologie). —

Die Quintessenz ist der ewig problematische Zusammenhang alles Gedanklichen mit dem Erlebbaren (Sinnen-Erlebnisse).

Die Arbeit für die Broglie-Jubiläums-Band wird von den dortigen Kollegen selbst ins Französische übersetzt. Der Inhalt ist für die Leute aber schlimmste Ketzerei. Ich kann es Ihnen erst schicken, wenn es gedruckt ist.

Uns geht es allen recht ordentlich. Mein Arbeitsvermögen hat aber schon empfindlich abgenommen. Nun, es hat auch sein Gutes. Ende alles Herzliche

Ihr A. E.

Figure 10.1. One of the last lessons of Einstein to Solovine, 7 May 1952. (© The Albert Einstein Archives, The Hebrew University of Jerusalem)

we go back to reality, where those predictions will be subject to observation, to experimentation (represented by the vertical arrows), in order to verify if there is agreement between the various S's and E's, that is, between prediction and observation.

Behind this diagram we can make out the structure of general relativity. (A) represents the theory itself, with its principles in the form of axioms and its entire mathematical structure. From the theory, we can deduce theorems and propositions, in the present case three, corresponding to relativity's three classical tests.

In this letter to Solovine and in the accompanying diagram, Einstein stresses the "jump," the arrow on the left going up from (E), experience, to (A), a jump that allows the theoretician to formulate "his" theory—a mathematized theory that includes the principles, concepts, mathematical tools and methods, but also the equations fitting them into a physical theory. This is the result of a creative act, a free act that the researcher has no obligation to justify a priori. That jump represents the instant at which and the creative gesture by which the theoretician invents his theory. There is more work to be done in this essentially mathematical space: first to properly understand it and make sure that it is not inconsistent from the mathematical point of view or contradictory on the physical level, and then to deduce observable consequences in the form of tests or effects. Later, returning to reality, the researcher will have to compare those propositions against the observations, in short, he will have to verify the theory. "To verify" is now a somewhat ambiguous term, because he no longer has to make sure that the theory is true in a strong sense, as a mathematical theorem would be, but, more humbly, that the theory is "right"; that it correctly accounts for the observations. That it is not less interesting that the theories that preceded it.

In order to illustrate the diagram for his friend, Einstein chose an example from daily life: "the concept of dog and the experiences attached to it."[50] A theory of the dog, actually, which he only outlined, but which I will develop quite freely in order to drive the point home. (E) is the dog itself, the dog as we experience it. (A) stands for the idea we have of a dog, our representation of it, the theory that we, as canine specialists have constructed—a theory (in general not expressed in mathematical terms!) that we have invented, of course, and from which we deduce certain propositions (it eats, it barks, it mates with other dogs and puppies are born). These propositions will then be used to verify that our theory not only explains the dog's behavior but is also able to predict it (S, S′, S″). This information will be compared to the actual behavior of a dog by means of canine experiments. If everything is all right—that is, if the propositions are verified—then we will have a theory of the dog or, more modestly put, a model of the dog. But if a proposition fails the test, we will have to modify the axioms and principles of the dog in (A) and start the cycle again, until we obtain a theory as precise as possible, above all a canine theory better than the previous one.

It would be useful, however, to formulate these questions in a more physical and less formal way and ask what the connection is between physical reality (E) and physical concepts (A), as regards, for example, the concept of curvature in the space of general relativity. It is out of the question, and useless, to wonder whether the space of the theory is *really* curved or what a curved time would look like. We simply have to accept the fact that the mathematical apparatus that incorporates the notion of curvature into a Riemannian space (in (A)) and on which the concepts and objects of the theory are based, accounts for the observed phenomena better than some other theory—Euclidean, for instance. That is essentially all there is to it. The idea of a curved space has no physical reality (it does not *really* exist in (E)); it cannot be directly observed. However, the bending of light rays really takes place and can be observed, but it is a different phenomenon that has nothing to do with the curvature of space, which is a representation with no physical reality. The *curved* character of general relativity's Riemannian space is no more meaningful a priori than the absolute character of Newton's space (or the theory of the dog!), and it is only justified a posteriori in the context of the global structure of the theory and because the latter is (or is not) confirmed by experience. In a nutshell, according to Einstein, the theory or theoretical representation lives in our thoughts, in our books, in our culture. It is, in his own words, "the eternally problematic relation of interdependence between the creations of the mind and reality as experienced through the senses."[51]

Above all, Einstein's analysis shows how and why, since its creation, general relativity had dominated the field of gravitation: because of its consistency, and especially because it accounted for all available observations better than any other theory. And yet, despite its success, Einstein's theory of gravitation remained an unpopular theory until its renewal in the 1960s. It was, perhaps, sometimes too popular among the relativists, of course, but more often it was despised, occasionally even abhorred, by the specialists in other fields of physics. But it was always the source of a reciprocal animosity that helps explain the stakes of a battle that the relativists, on their geometric pedestal, were so certain of having won, while the others hoped for an end to general relativity's arguable but inescapable domination.

NOTES

1. This chapter as well as the next one are largely based on the author's article Eisenstaedt, 1986, in which the reader will find a more detailed analysis.
2. Lanczos, 1932, p. 113.
3. Bergmann, 1942, p. 211. My italics.
4. Weyl, 1918, p. v.
5. Einstein, 1929, p. 114.
6. Ibid.
7. E. Rutherford, quoted in Chandrasekhar, 1979, p. 213.
8. E. Rutherford to G. E. Hale, 13 January 1920, quoted in Crelinsten, 1984, p. 50.
9. Crelinsten, 1981, p. 160.
10. Quoted by A. Sommerfeld in Sommerfeld, 1949, vol. 1, p. 102.
11. O. Lodge, 1919, quoted in Crelinsten, 1980, p. 189.
12. J. J. Thomson, quoted in Chandrasekhar, 1979, p. 213.
13. Thomson, 1920, p. 560.
14. Bouasse, 1923, p. 8. Henri Bouasse was then professor of physics at the Faculty of Sciences at Toulouse.
15. Stachel, 1974, p. 32.
16. Einstein, 1949a, p.75.
17. C. Lanczos to G. J. Whitrow, in Whitrow, 1967, p. 49.
18. Stachel, 1974, p. 32.
19. Einstein, 1921a, p. 784. His second pole of interest is cosmology.
20. Einstein, 1930, p. 1.
21. Pauli, quoted in Pais, 1982, p. 347.

22. Quoted in Pais, 1982, p. 273.

23. Einstein, 1942, in Bergmann, 1942, p. iii.

24. On the difficulties of this program, cf. Goenner, 1984.

25. This is for instance G. C. McVittie's view, McVittie, 1956, p. 2.

26. Lanczos, 1955, p. 1213. Lanczos is referring here to the ten years that separate the two theories of relativity.

27. Einstein to Born, 1949, in Born, 1969, p. 244.

28. Born, 1955, in Mercier and Kervaire, 1956, p. vi.

29. For example, H. A. Kramers, Q. Majorana, W. Pauli, and E. Schrödinger.

30. In this regard, cf. Bourdieu, 1976.

31. Pauli, 1958.

32. Ibid., p. vi.

33. Ibid.

34. Lanczos, 1932, p. 115.

35. Bondi, 1962, p. 325.

36. Infeld, 1950, pp. 69–70.

37. Wheeler, 1957, p. 4.

38. B. K. Harrison et al., 1965, p. 124.

39. Whitehead, 1922, p. vi. Regarding Whitehead, cf. also Chandrasekhar, 1979.

40. Milne, 1940, p. 52.

41. Milne, 1935, quoted by Chandrasekhar, 1979, p. 214.

42. Ibid.

43. Fouchy, 1751, p. 103.

44. Einstein, 1919b, p. 1.

45. Einstein, 1922, p. 429.

46. In particular in his treatise *The Logic of Scientific Discovery*, published (in German) in 1935.

47. For example, when interviewed by G. J. Whitrow; Whitrow, 1967.

48. Einstein, 1954, p. 282.

49. In a letter to his lifelong friend, Maurice Solovine, who was living in Paris and working at Gauthier-Villars, where he edited, among others, Einstein's works, which he translated into French. This 7 May 1952 letter has been analyzed in detail by G. Holton, in Holton, 1981, pp. 224–271.

50. Einstein to Solovine, 7 May 1952, in Einstein, 1956, p.121.

51. Ibid.

CHAPTER ELEVEN

An Unpopular Theory

THE YEAR 1919 was an auspicious year, when the English expedition wrote a decisive page in the history of the theory. General relativity was verified! The future appeared glorious. All that remained to be done was to patiently develop new consequences that would then be cautiously tested against observations. The relativists could rightly expect to see in the coming years a continuous and increasing stream of works aimed at developing the theory, consolidating its foundations, determining its connections with reality, and promoting its study.

That would be the story of a *normal* theory. But historians of science know well that such a simple scenario is never true. If we objectively consider its results and its central ideas, everything was going well, even very well, for Einstein's theory of gravitation. Wasn't it a winning theory? Not really, as we have seen. Far from a reasonable development, a sensible dynamics, and an unclouded history, general relativity went through a bumpy evolution and a slow and difficult growth, and the relativists had to endure isolation and doubt, if not a certain marginality.

During this period, the political situation certainly had a decisive influence on scientific production. General relativity was born with World War I, and so its development was clearly hindered. In the following years, Nazi domination, which especially targeted relativity—a symbol of "Jewish physics"—resulted in the flight or death of many scientists throughout Europe, and in particular the departure of Einstein for Princeton in 1933. World War II would later completely halt nonmilitary scientific production for more than five years.

By all accounts, the teaching of general relativity was greatly affected by the situation. If Einstein was occasionally concerned by this, he does not seem to have tried to do much about it in the

United States, because the first course in general relativity was only given in 1952 at Princeton University, and between 1936 and 1961, no course on Einstein's theory was offered at the University of Chicago. In fact, the teaching of the theory was left for each university lecturer to decide according to his or her own preferences. Born taught a course in 1929 at Göttingen and another one at Edinburgh in 1940. In France, a few courses were given at the beginning of the 1920s, by Jean Becquerel in Paris and by E. Bauer and E. M. Lemeray elsewhere in the country. But there is no evidence of any systematic teaching of the theory anywhere before the 1950s.

Leopold Infeld, a close collaborator of Einstein's, described the atmosphere prevailing at Princeton's Institute for Advanced Study, to which Einstein was appointed in 1933:

> In any case, the greatest interest in this discipline was evinced by scientists in the 1920s. Then, already in 1936 when I was in contact with Einstein in Princeton, I observed that this interest had almost completely lapsed. The number of physicists working in this field in Princeton could be counted on the fingers of one hand. I remember that very few of us met in the late Professor H. P. Robertson's room and then even those meetings ceased. We, who worked in this field, were looked upon rather askance by other physicists. Einstein himself often remarked to me: 'In Princeton they regard me as an old fool.' This situation remained almost unchanged up to Einstein's death. Relativity Theory was not very highly estimated in the 'West' and frowned upon in the 'East.'[1]

Peter Bergmann, another of Einstein's very close collaborators, observed that in those years "you only had to know what your six best friends were doing and you would know what was happening in general relativity."[2] It is a testimony that underscores the lack of communication among relativists, isolated in their universities, but which should not be taken at face value, for it would restrict the discipline's specialists to a few faithful close to Einstein.

Einstein did not play a rallying role; his interest in his general theory of relativity was erratic. He had very little clout with the Institute's administration and did not succeed in obtaining a position at Princeton for Infeld, who settled in Canada before returning to Poland. In 1949, Einstein wrote to Born explaining the failure of his efforts to procure him an invitation to the Institute: "I have little influence, because I am considered here as a kind of fossil who went blind and deaf over the years. I do not find that role at all unpleasant since it suits quite well my disposition."[3]

The relativists were isolated. Each one pursued his own line of work in the context of an interpretation of the theory that he did not always share with his colleagues. In fact, the research structures in general relativity remained rudimentary, cut off from the other main trends in physics and scattered in the shadow of universities where Einstein's theory was studied mainly out of personal interest or curiosity or for the simple pleasure of understanding it better.

The vastness of the theoretical field—the scattering of specialists and journals in numerous branches and different countries—did not facilitate communication or collaboration. The legitimacy of the subject was not easy to impose, even if Einstein's authority, the impact of some works, and the interest of certain questions occasionally permitted it. The renewal of the discipline in the 1960s had its roots partly in a number of studies whose interest was not always immediately appreciated but thanks to which a number of open theoretical questions could be satisfactorily answered. But, insofar as there were no urgent demands from the observational front and no experiment crystallized the theoretical questions, general relativity, as a specialty, was at a standstill. At the institutional level, its lack of dynamism resulted in an absence of positions and funds. As a consequence of that, fewer results were generated as interest in the theory waned. The striking shortage of specialists, who where spread over too large a front during those years, was the reason—and the consequence—of the theory's stagnation. In short, the critical mass that would allow the theory to be sustainable was far from being achieved, and the theory's revitalization had to wait for the replenishment of its specialists.

At the end of the 1950s, the concern of general relativity's specialists was at its peak, as illustrated by this passage from a technical text by Andrzej Trautman, a well-known Polish relativist, that appeared in a very influential Russian journal: "Many physicists are very emotional about the Einstein theory. Most of them admit that the general theory of relativity is a beautiful theory, but then they add that because of the weakness of gravitational forces the theory of gravitation is on the sidelines relative to the rest of physics. . . . Most discussions begin or end with the reproach that GR has been subjected to too few tests—as if this were a fault of the relativists."[4] After having complained that "some physicists go as far as to say that GR should not be regarded as a physical theory," he mentions those who consider the relativist as "mathematicians rather than physicists," before lambasting "certain circles" in which relativists are regarded as "socially undesirable elements," as well as "certain relativists" who "regard Einstein's theory as standing in some higher relation to other theories."[5] In a postface, he concludes with a decidedly ecumenical stance: "If there are mathematicians who want to study the equations of a physical theory, they should be welcomed, and not subjected to ostracism by a refusal to regard their work as part of physics."[6]

From the beginning of the 1920s to the end of the 1950s, we have seen that the situation of general relativity deteriorated, having an impact on the hiring of relativists, the interest in teaching the theory, the volume of publications, and the place of the specialty in international conferences. The decline of the theory also affected the morale of the relativists.

The diagram in figure 11.1 shows, for each year, the number of publications (according to the specialized journal *Science Abstracts*) in the field of general relativity from its birth to the end of the 1950s. We can see the spectacular growth of the theory after its birth, followed by its peak in 1920 (the verification of the second test took place in 1919). The fact that the theory became fashionable at the beginning of the 1920s is reflected in the increased number of publications. And then there is a sharp and catastrophic decline starting in 1922–23, a situation that lasted until the end of the 1950s. There is, however, a small rise that indicates a slight improvement around the 1930s and corresponds to the work prompted by the interest in cosmology. General relativity thus experienced its "crossing of the desert."

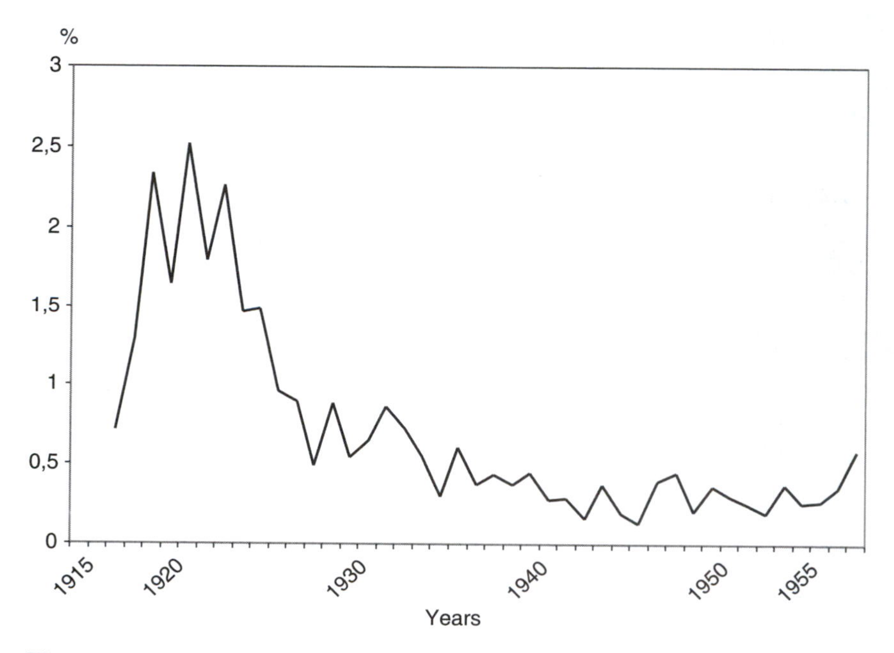

Figure 11.1. Number of articles in general relativity as a percentage of the total number of publications in physics from 1915 to 1955.

From the beginning of the 1920s—general relativity's golden years, when Einstein's fame was established—to the mid 1930s, Newtonian mechanics held its position as regards the number of publications in physics (7 percent), while general relativity's share declined from 7 to 2 percent. Quantum mechanics, which only did half as well as Einstein's theory in 1920, would turn the situation around ten years later.

THE BERN CONFERENCE

The time is July 1955. Einstein has just passed away in Princeton, and in Bern, Switzerland, a celebration, "Fifty Years of Relativity Theory" (special relativity), is taking place, precisely in the city where the theory was created. But, ironically, it is essentially to general relativity, whose fortieth anniversary is discreetly ignored, that the conference will be devoted. Although most of the

Figure 11.2. —Quel concert de louanges! Ces dames parlent encore de leurs couturiers. —Mais non, il s'agit d'Einstein. (—What a barrage of praises! These ladies are still talking about their couturiers. —Absolutely not. They're talking about Einstein.)
Drawing by P. Portelette. *L'écho de Paris*, 8 April 1922. (© Charmet Archives)

talks are of a technical nature, Einstein's death provides an opportunity to pay him tribute, a tribute that, inevitably, takes at times the form of hagiography. Max Born's speech is for this reason even more interesting and memorable, especially coming from Einstein's closest friend and colleague among those attending the conference.

Born is one of the last to speak, on Friday afternoon, just before the conference's closing. He talks mostly about the history of special relativity, a theory of tremendous importance for physics in general and for quantum physics, his specialty, in particular. He refers to general relativity only once, when he evokes the moment he met Einstein and how he learned about general relativity. Let us listen to him:

> I remember that on my honeymoon in 1913 I had in my luggage some reprints of Einstein's papers which absorbed my attention for hours, much to the annoyance of my bride. These papers seemed to me fascinating, but difficult and almost frightening. When I met Einstein in Berlin in 1915 the theory was much improved and crowned by the explanation of the anomaly of the perihelion of Mercury, discovered by Leverrier. I learned it not only from the publications but from numerous discussions with Einstein—which had the effect that I decided never to attempt any work in this field. The foundation of general relativity appeared to me then, and it still does, the greatest feat of human thinking about Nature, the most amazing combination of philosophical penetration, physical intuition and mathematical skill. But its connections with experience were slender. It appealed to me like a great work of art, to be enjoyed and admired from a distance.[7]

If Born can speak in those terms, so candidly, and during a conference, it is because for forty years he had been Einstein's friend. Their correspondence was one of the first to be published, and it shows to what extent both men—both couples, in fact—were close. They first met during the summer of 1915 in Berlin, where Born was scientific advisor to the commission for artillery control. Einstein, who had just separated from his first wife, Mileva, and his two young sons, had been appointed a member of the Prussian Academy of Sciences and had started a new life with his cousin Elsa. The two men got along perfectly well, not only on scientific issues but also on political and human questions in general. It is therefore this intimacy that allows the old friend to say what he really thinks of general relativity, on which, in spite of what he says, he has published several papers. In short, nobody is in a better position than Born to assess the interest of the theory for the physicists of his time. He has probably never worked on the subject from a technical point of view, but isn't it precisely this distance that authorizes him to take such an uncompromising look at the place of general relativity?

The scene inside the lecture hall of Bern's Museum of Natural History is quite extraordinary. As Born speaks, we have to imagine the amusing stupefaction of certain of his very dear colleagues, stifling in the July heat, at being offered at that rather solemn moment of the conference the reminiscence of an intimate episode of his life, his honeymoon, from which he took time off to read that "frightening" work. It is also for Born a way of expressing his interest in those ideas and their author. An interest that is immediately contradicted by the declaration that follows. In a few words, he gives the reasons why he decided "never to attempt any work in this field": most likely because the theory appeared to him difficult and even frightening. But the crucial point of this analysis of his reasons is the shortest sentence, the guiding principle of every physicist: experience. "But its connections with experience were slender"—that is for Born the real problem with this theory, the reason he admired it "from a distance," which does not prevent him from also expressing, with genuine conviction, the reasons for his admiration, which he shares with many of his colleagues.

It was therefore because of its scant connections with the "real" problems in physics that Einstein's theory was for a long time rejected as merely a piece of art for art's sake. Did three little tests, which were still being challenged, justify that impressive mathematical arsenal? What was the use of general relativity? The unspoken question was on many minds, if not on those of concerned relativists.

At the end of the 1950s, far from the radical nature of the vision of space-time it proposed, general relativity was no longer considered a *revolutionary* theory; it had lost that epithet to quantum theory, and it would try to win it back through its renewal. The new generation of wave mechanics theoreticians would not show as much understanding or sympathy toward the theory as Pauli and Born had. Richard Feynman, invited at the turn of the 1960s to the first conferences on gravitation, would be infinitely less friendly, slighting his colleagues in a recently published letter to his wife: "I am not getting anything out of the meeting. I am learning nothing." "It is not good for my blood pressure." Let us skip his almost abusive remarks and focus on his analysis: "Because there are no experiments this field is not an active one, so few of the best men are doing work in it." "It is not that the subject is hard; it is that the good men are occupied elsewhere. Remind me not to come to any more gravity conferences!"[8] Such comments were unacceptable to the relativists, who felt rejected by their quantum colleagues.

THE IVORY TOWER

Even if the specificity of relativity hinged on technical, epistemological, and institutional factors, there was actually more to it than that. If there were so many reasons to work in quantum mechanics, and Born gave us a few of those, what was the motivation of those who would prefer to bury themselves in the study of general relativity? In the preface to his *Relativity: the General Theory*, a fundamental work published in 1960 and which more than any other represents the totality of the work carried out throughout those years, this is the question John Synge asks, as he describes, with a good sense of humor, the image he has of . . . himself:

> Of all physicists, the general relativist has the least social commitment. He is the great specialist in gravitational theory, and gravitation is socially significant, but he is not consulted in the building of a tower, a bridge, a ship, or an aeroplane, and even the astronauts can do without him until they start wondering which ether their signals travel in.
>
> Splitting hairs in an ivory tower is not to everyone's taste, and no doubt many a relativist looks forward to the day when governments will seek his opinion on important questions. But what does "important" mean? Science has a dual aim, to understand nature and to conquer nature, but in the intellectual life of man surely it is the understanding which is the more important. Then let the relativist rejoice in the ivory tower where he has peace to seek understanding of Einstein's theory as long as the busy world is satisfied to do its jobs without him.[9]

Beyond its ironic tone and provocative character, there is no question that this text is representative of most of the relativistic world and its stakes. Here, Synge schematizes the opposition between two worlds, two cultures that until then had few points in common other than the sharing of a same ambition and a same institution: theoretical physics. These are two almost airtight worlds, made up of specialists who most of the time ignore each other, working in disjoint fields, embracing opposing philosophies, and pursuing irreconcilable goals: to understand or to conquer. On the one hand, the specialists of the infinitely small look condescendingly at those theoreticians sitting on their geometric pedestal, pure thinkers who can boast a few seconds of arc but whose theory shines on the pediment of philosophical journals and mundane conferences with no relation to its real interest. To that attitude the relativists respond from the safety of their ivory tower, a shelter they badly need to protect themselves from that inferiority complex which Synge, in another text, exhorts them to overcome. It is above all a primitive, even monastic conception of science and the scientists that he defends here: the rejection of both a triumphalist physics and an alienating society, not unlike the position of Einstein himself; it is the skepticism of an idealist who does not expect much from the evolution of modern science.

But beyond his personal views, the questions troubling the relativists are clearly presented. These are rooted in the logical connections between theory and experience, between physics and mathematics, as we have observed above, but even more so in the almost industrial organization of the world of physics, more than ever tied to economic, military, and political powers—a world that so far the relativists have managed to avoid but with which (Einstein forbid!) they are increasingly confronted, because of the proximity in the universities of quantum and theoretical physics and because of those who—coming from, or tempted by, that world—would like to organize the discipline according to its methods.

Of course, not everybody agreed with Synge's choices, but the essence of his analysis can certainly be found here and there. A few years later, on the evening of the first Texas meeting—a gathering devoted to a booming specialty, relativistic astrophysics—a first-class speaker, T. Gold (wasn't he invited to give the traditional after-dinner speech?) would express, in no uncertain terms, the opinion of many of his colleagues, and not only those from the new generation, by rejoicing at the fact that "the relativists with their sophisticated work were not only magnificent cultural ornaments but might actually be useful to science!"[10]

Synge's text marks particularly well the boundary between two periods, separated symbolically by Einstein's death and more officially by the Bern conference: a first period, during which general relativity remained, within theoretical physics, a somewhat outdated haven, protected from the strong currents agitating the quantum theories; and a second period, the time for renewal that Synge so dreaded. It was a boundary that the organizers of a 1973 school on *astres occlus*, or black holes, marked in the introduction to the proceedings: "The story of the phenomenal transformation of general relativity within little more than a decade, from a quiet backwater of research, harboring a handful of theorists, to a blooming outpost attracting increasing numbers of highly talented young people as well as heavy investment in experiments, is by now familiar."[11]

In short, relativists would from now on be able to live off general relativity, and no longer only for Einstein's theory. The renewal had begun.

NOTES

1. Infeld, 1964, p. xv.
2. Peter Bergmann, Einstein's student in the 1930s and later one of the best-known relativists after the war. In 1942, he wrote a book on relativity with a foreword by Einstein (Bergmann, 1942). Quoted by Pais, 1982, p. 268.
3. Einstein to Born, 12 April 1949, in Born, 1969, p. 245.
4. Trautman, 1966, p. 334.
5. Ibid.
6. Trautman, 1966, p. 334.
7. Born, 1956, p. 253.
8. Feynman et al., 1995, p. xxvii.
9. Synge, 1960, p. vii.
10. Gold, 1965, p. 470.
11. De Witt and De Witt, 1973, foreword, p. vii.

CHAPTER TWELVE

The Rejection of Black Holes

WE ALL HAVE heard of black holes. Countless popularization articles have been devoted to the subject, and every year a news story suggests that this time it is true, some astrophysician "has [finally] discovered one." And it is always the first time. Has anybody really seen a black hole? We shall return to them in chapter 14, but not before examining the question: what is a black hole?

Nothing is more important for a physical theory than to predict, and later discover, a new particle, a new element, a new planet, or a new type of star. A black hole is in fact a new kind of star, a celestial object totally different from anything previously known—an object that Einstein's theory predicts, but to which no relativist, including Einstein himself, had given a thought before the 1960s. The idea of a black hole was not contemplated when the theory was born. It was only recently, toward the end of the 1960s, that the concept was created, and it was understood and accepted after the 1980s, more than fifty years after the birth of the theory on which it depends. It is a notion that raised some fundamental questions and required a practically new interpretation of general relativity. The emergence of the concept of black hole was nothing less than the second relativistic revolution.

And yet, from the moment general relativity was formulated, a certain characteristic, a certain *singularity* of the theory, posed problems that would be solved only fifty years later. These questions that the relativists, Einstein in the first place, faced throughout the century are still our questions today, at least in part. For this reason it will be useful to retrace the history of this dark object, in order to make sure that those questions have received acceptable and convincing answers. In the spirit of the philosophy of this book, the present chapter will be partly historical and partly theoretical.

In the next two chapters I will explore the current notion of the black hole and the way in which the concept evolved after the 1960s. I will also discuss the contemporary interpretation of this strange object from two different points of view: First, from a theoretical perspective—What is the meaning of this region of space as a geometrical object?—because space-time is involved here, the space-time generated by a collapsed star. Then from a physical standpoint—What is the nature of this weird star and how can it be created? What is left of the star itself? What place does it occupy in today's astrophysics? And, finally, do black holes exist? I will then investigate whether, and to what extent, they were an inevitable consequence of the theory, that is, up to what point general relativity really predicted the existence of such a controversial object.

Black holes took form not only within the framework of the equations of general relativity and astrophysics, but also in the minds of the specialists, specialists who revolutionized their theory while others resisted almost to the point of alienating themselves from their community. The answers proposed by those new-wave relativists were stunning. They advanced a new world, a strange space-time, far removed not only from Newton's absolute space but even from that of special relativity. The resistance of old-time relativists was therefore not surprising, and it was probably needed, so that all the essential questions could be asked and this incredible and strange reinterpretation not be accepted too hastily. It was also necessary to be firmly convinced that this second relativistic revolution was inevitable—inevitable but far from obvious. We must keep in mind that the concept of black hole is not an easy one, and that it is probably as difficult for an ordinary person to accept it today as it was for the relativists back then.

SCHWARZSCHILD'S SINGULARITY[1]

Karl Schwarzschild, a German astrophysicist, had long been interested in these questions. Not only did he closely follow Einstein's

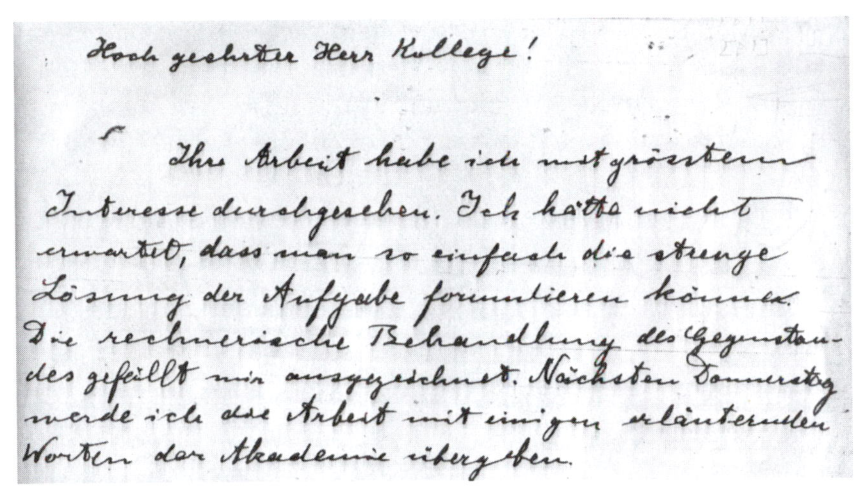

Hochgeehrter Herr Kollege!

Ihre Arbeit habe ich mit grösstem
Interesse durchgesehen. Ich hätte nicht
erwartet, dass man so einfach die strenge
Lösung der Aufgabe formulieren könne.
Die rechnerische Behandlung des Gegenstan-
des gefällt mir ausgezeichnet. Nächsten Donnerstag
werde ich die Arbeit mit einigen erläuternden
Worten der Akademie übergeben.

Figure 12.1. "Highly esteemed colleague, I examined your paper with great interest. I would not have expected that the exact solution to the problem could be formulated so simply. The mathematical treatment of the subject appeals to me exceedingly. Next Thursday I am going to deliver the paper before the Academy with a few words of explanation." A. Einstein to K. Schwarzschild, 9 January 1916. (German original in *Collected Papers of Albert Einstein* [*CPAE*], vol. 8, p. 239; English translation in *CPAE*, vol. 8, p. 175)

work on gravitation, but he had, in a way, preceded him because at the very beginning of the century he had already envisaged, in keeping with the nineteenth century works on non-Euclidean spaces, the possibility of a curved universe—whose structure could be spherical, elliptical, or hyperbolical—and had calculated its radius.

Einstein had just obtained his field equations for empty space and then set out to calculate the approximate form of the gravitational field around a spherical star. His was a typical neo-Newtonian interpretation, where the Sun's gravitational field was hardly different from the one considered correct by Laplace, but it was enough to solve the problem of the advance of Mercury's perihelion. Schwarzschild tackled the same problem from a more theoretical angle and succeeded in solving—exactly—the problem of the structure of space-time in the vicinity of a spherical distribution of mass. This was the first exact solution of general relativity, and it was therefore named after him. Schwarzschild's solution was extremely important in the history of the theory, for it gave rise to some fundamental problems. Let us see which ones.

Schwarzschild soon realized that his *exact* solution was not the same as the approximate one Einstein had just published; far from it. Near the center of space-time, Schwarzschild's solution had an odd behavior: the roles of time and space are reversed; time takes the role of space which takes the role of time. On this sphere, soon to be named after its creator, at the moment this reversal of roles takes place, everything becomes strange, weird. Proper distance is no longer well defined, because certain mathematical coefficients needed to calculate it become zero or even infinity. But, at that place, what does space-time itself become if its definition doesn't make any sense?

Schwarzschild took an interesting step: he assumed that space-time only began outside this sphere and ignored everything that caused a problem. Some specialists would half-heartedly object, because they saw no reason for that particular hypothesis. In fact, nobody knew what to do about the problem. The relativists, who then numbered only a handful, could have rejoiced and declared: "Finally space-time is no longer space plus time; something is going on in general relativity's kingdom!" But they did nothing of the sort; they denied everything. And, for a long time, the interior of the sphere was a no man's land. Almost fifty years passed before anyone knew what to do with that strange and forbidden place on which something new would be built, something we today call a black hole.

Nobody understood the meaning of that sudden turnabout of space and time, and because nobody understood it, there was an impasse. The peculiar character of that strange place was for them an excellent reason to stop there, on this singular sphere—on this *magic sphere*, as Eddington called it—where light seemed to come to a stop. However, this weird region was not a cause for concern, given that it was a virtual place that would probably never materialize, it was hoped, in the real world. In fact,

Schwarzschild's solution represented the space *outside*, the empty space around the star, and therefore it did not describe the internal structure of the star itself. And the radius a of the star was, in general, infinitely larger than that of this singularity (whose radius was $2GM/c^2$; see box 5 in chapter 7). Thus, Schwarzschild's singularity remained a virtual one, hidden at the center of the star, unable to manifest itself, or so it seemed—except if one assumed that the radius of the star was smaller than the radius of the singularity; in other words, that the star has disappeared behind this boundary of the world. It was hard to imagine such a situation, for no stars of this kind were then known, of course, and it was believed that none could exist. Much later the following question would arise: do stars exist that are so dense they can disappear behind this singularity? It would take some drastic conditions for that to be the case, and we will discuss them in detail later, but a simple calculation can provide an idea of what is involved here: the Sun's Schwarzschild's radius ($2GM/c^2$) is only 3 km, whereas the Sun's actual radius is 696,000 km. For the Sun to exhibit such a behavior, that is, for Schwarzschild's *singularity* to manifest itself—in other words, for the Sun to become a black hole—its radius would have to be smaller than 3 km. At the time, actually not so long ago, it was unimaginable that such a dense or massive star could exist for the phenomenon to occur.

So where's the problem? Why worry about an inaccessible place? Why worry over something that would never happen? Before following Einstein's colleagues along this path, we must realize that, in any case, it is important to understand and to study each object, each concept of the theory (and even more so the strange ones), and each one of its solutions to have an idea of where it may lead. This is necessary, even crucial, if we truly want to understand how the theory is actually built. For, who knows, the theory might turn out to be inconsistent. Therefore, even if one believes that "that will never happen," it is useful and interesting (and amusing) to study the *whole* of Schwarzschild's solution over the *entire* interval of definition of the variables: assuming that its source (the matter at the origin of the gravitational field) is not an ordinary star but is extremely dense, reduced to a point located at the very center of space-time, where $r = 0$. The question we should study is, therefore, the structure of the entire space-time, from A to Z.

The meeting that took place in Paris in 1922 is a good example of the questions surrounding this issue and of Einstein's reaction. This was Einstein's first visit to Paris, four years after the end of WWI, following an invitation from Paul Painlevé, an excellent mathematician but a poor relativist and a former war minister, and from Paul Langevin, one of the best French physicists of the time, a man with a passion for relativity. Hadn't he invented the paradox of those twins whose ages are different after having traveled their separate ways in the space of special relativity? Langevin was also Einstein's friend, a leftist who worked for peace, as Einstein did.

The problems stemming from the infinite character of the coefficients of Schwarzschild's solution—the gravitational potentials, as they were then called—were at the center of the discussions. Beside Einstein, Painlevé, and Langevin, there were Jean Becquerel, Marcel Brillouin, Élie Cartan, Théophile de Donder, Jacques Hadamard, and Charles Nordmann among the large crowd filling the physics hall of the College de France.

Hadamard, a French mathematician who specialized in differential geometry, was truly concerned over Schwarzschild's singularity. What was going on at that place? Einstein himself was embarrassed for, if that term was really zero or infinity—in other words, if it was singular—it would be disastrous, an "unimaginable tragedy" for the theory. And what would happen from a physical point of view? A tragedy that Einstein called, amusingly, the "Hadamard catastrophe."[2] However, as far as they knew, astrophysics still being in its infancy, there seemed to exist no star whose mass would be much greater than the Sun's or whose radius would be smaller than its Schwarzschild radius. It was in the nature of things, they believed, that there should be an insurmountable limit to the growth in a star's mass that would prevent celestial bodies much larger in mass than the Sun from existing; this limit would preclude the appearance of the "Hadamard catastrophe": it would be a truly forbidden zone.

Such speculations were not enough to reassure Einstein, who at the meeting's next session presented a little calculation according to which, during the collapse of a very large star, a *physical* catastrophe, due to the star's internal pressure becoming infinite, would precede the "Hadamard catastrophe." It was as if, unable to imagine a geometric catastrophe, they preferred to bet on a physical one. Hadamard seemed satisfied and believed the much feared catastrophe to be impossible. In fact, all these prognostications were extremely shaky.

The interpretation favored at the Paris meeting was typical of the general sentiment that prevailed until at least the 1950s. For the relativists, space-time was, ultimately, given a priori and it was still essentially Newton's absolute space, at least from a spatial point of view. It was unthinkable, inconceivable (except in cosmology, as we shall see later), that space could actually be locally curved, here and now, except very slightly. Gravitation was a weak interaction acting only at long range and therefore unable to strongly modify the local geometric structures. The fact that space-time might be subject to a true dynamics, that its topological structure, its shape, might be strongly modified by an accumulation of matter was simply not conceivable. The interpretation of the theory remained essentially Newtonian in spirit: it was a neo-Newtonian interpretation.

EXPLORING SCHWARZSCHILD'S SPACE

The study of the trajectories (that is, the geodesics) of a given space-time of the theory is extremely interesting since, by describing and following all possible and imaginable trajectories, we will be able to know what is really going on, what the structure of this particular space-time is. Of special interest is the study of the trajectories of light for they are, in a way, a limit case: being the trajectories of the fastest particles they are the least affected by gravitation.

There are certain similarities here with the discovery of the spherical nature of the Earth that also reveal the difficulties encountered by our relativists. Let us travel back in time to Greece, to the sixth century before Jesus Christ, at the time of Thales of Miletus, who is believed to have been the first to suggest that the Earth is round. The idea of a flat, disk-shaped Earth surrounded by the "river" Ocean then prevailed—an image born out of reality as perceived and rooted in certain myths. If we recall how difficult it was for us to accept this fact in our childhood, we may better understand the difficulties Thales faced in conceiving this image, accepting it as inevitable, and trying to persuade his contemporaries. It took the full weight of Socrates' and Plato's authority in the fifth century, and especially Aristotle's in the fourth, for people to accept that the Earth is a sphere, immobile of course, around which the Moon, the Sun, and the fixed stars revolve. What a tremendous intellectual effort was required to imagine it and to believe it! The facts were there: the Sun revolving around the motionless Earth, around the Moon, the existence of the horizon; but they were up against a physical ideology that had prevailed since the birth of civilization. The transition from an image of the world built around a disk to a conception involving a sphere was immensely arduous—not to mention picturing the situation at the antipodes: why don't things fall off? What a difficult idea for human subjectivity to accept! Astronomical observations and, centuries later, journeys across the oceans would shatter the stable image of the world built in harmony with the familiar scenes: village, Sun, or fixed stars.

This great moment, the discovery that the Earth is a sphere, may help us better understand the reaction of our relativists faced with the explosion of space-time and the efforts they had to make in order to conceive the relativistic revolution. To begin with, one of the best ways to understand space (time) is to explore its paths. We can actually, or in thought, travel the paths on the Earth, and we can explore Schwarzschild's space thanks to his equations. But our exploration of the universe and its limits need not be a real journey (although some revolutionary progress in the theory was made in the 1930s thanks to the powerful telescope at Mount Palomar) but only a mental one. By exploring the trajectories of Schwarzschild's solution we will be able to acquaint ourselves with the theory and its limits and with the kind of space-time it has in store—a complete space, for it must include all possible trajectories.

The study of *all* trajectories of Schwarzschild's solution is a good way to discover the space-time created by a star and to get to know its boundaries, its shape, and its topology. This is one of the few ways, one of the best ways in fact, to describe space-time; but we should not let our a priori idea of the journey prevent us from moving forward. For this is precisely what happened during those classical—too classical—years, when the relativists still believed that space was a lot like that of the ancients, that is, Newton's space. We only discover what we set out, in our mind, to discover; be it the *representation* of the Earth, as they said in the eighteenth century, or the structure of space-time.

Here again, the question of the representation of the Earth is enlightening. Suppose that the members of a certain (hypothetical) tribe believe that the Earth is flat; they cannot conceive that by traveling west far enough they would arrive at their starting point. They would probably never attempt the trip, simply because they don't see the point. One day, after a long journey west and by some incredible coincidence, a more intrepid traveler finds herself back at the place from where she left. This first explorer has little hope of being believed, but when other travelers report similar experiences, the fact begins to sink in. It will take a long time to realize that it can be explained by recognizing that they live on the surface of a sphere. That's a funny idea, come to think of it!—one that demands a tremendous change in mentality and considerable intellectual work. The observation of the sky—the Sun's daily reappearance and the Moon's partial eclipses—could confirm those ideas and bring people to accept that the Earth is spherical. And different points of view can certainly help to convince and be convinced, for it is not enough to account for those facts within a novel view of the world, but also, and foremost, one must convince oneself and then convince one's contemporaries of the truth of this new reality. Something (a little bit) like this happened to the members of the relativistic tribe.

The relativists traveled, in thought, across Schwarzschild's space, exploring the theory, seeking to discover its paths through the study of the so-called geodesics, the possible trajectories, but without really looking for America, without imagination! They believed they already knew the global structure of space-time, which could only be that of Minkowski's space, possibly slightly warped. The question of its topology, of the shape of space (does it resemble a plane, a sphere, a cylinder, or a torus?), only dawned gradually, within the small circle of cosmologists. They tried first to recover the classical trajectories and determine which changes general relativity made to Kepler's ellipses. They performed countless calculations and published numerous articles, monographs, and theses aimed at writing down the equations of the trajectories arising from Schwarzschild's solution and solving them and classifying them; but they did not attempt to map out Schwarzschild's space-time, which they thought they knew from the start. The mathematical questions were easily resolved, for the necessary techniques to integrate the equations and exactly determine those trajectories were already available. A large number of works very thoroughly covered all (or almost all) possible cases but without tackling the essential questions.

The reason is that the relativists were convinced in advance that Schwarzschild's space resembled Newton's, at least from a spatial point of view: Euclidean space plus time. They realized that the spatial component might be slightly bent, that time may be a bit out of step, but nobody imagined that Schwarzschild's solution could represent a space *really* different, completely different, from Newton's. For this solution represented, let us not forget, the geometry of space-time in the vicinity of a star, and it was not easy then, any more than it is today, to imagine that such a space-time could be twisted, despite the fact that torsion does not exist in general relativity. It was easier to imagine and accept a curved universe on a large scale, to acknowledge that its shape could be different from the Euclidean structure, perhaps as a consequence of the works of the previous century on non-Euclidean geometries. The cosmologists would be the first to reflect on these geometric concepts and topological questions. But it was unthinkable, before becoming incredible and later unbearable (the relativistic community went through all those feelings), that space-time could be warped, twisted, torn in its local structure, here and now.

Obviously, and predictably, to study this space they had to resort to tools that were almost Newtonian: to begin with, coordinate time t, which, just as Newton's absolute time, ranged over (practically) all space and its trajectories. This was already a serious mistake (but how could they not make such a mistake when nothing forced them to proceed otherwise!) for they chose to ignore something they already knew perfectly well in special relativity: that they should use physical time—proper time—and more precisely the proper time of each particle, which takes each planet along its own path. But, for technical reasons, they preferred to use the coordinate time t appearing in the already classical Schwarzschild's solution, a time in the image of Newton's absolute time: universal and everywhere defined—well, almost everywhere, except on and inside Schwarzschild's sphere. The same thing happened with the space variables: they de facto favored the classical coordinates used in the original solution of the equation. This was not too serious in the case of the angular variables, but it was catastrophic for the radial coordinate r, which was physically interpreted in the classical way, as the distance to the center of the star, despite numerous warnings from Einstein himself.

What was at play here was general covariance, a point over which everybody stumbled, and at times even Einstein. General covariance demands that *no* coordinate be favored, no coordinate should have a physical meaning, and this certainly includes Schwarzschild's. Conversely, this also implies that *any* possible coordinate system may be used and that *none* of them has a priori any physical meaning. But, general covariance notwithstanding, it was so tempting to hold onto the classical coordinates (r and t, so similar to the Newtonian coordinates) that worked so well, so smoothly, because the use of classical coordinates in their calculations nevertheless enabled them to find everything they set out to find. This approach was not particularly clear or neat, nor particularly precise, either, but it worked. And so they recovered all classical Newtonian (Keplerian) results, and in like fashion they analyzed the theory's three classical tests that were, after all, only minor deviations from the Newtonian rule. So why complicate things? They ended up by devising a neo-Newtonian interpretation not only of Schwarzschild's solution but of practically all problems in general relativity.

Of course, all those results were useless when applied to Schwarzschild's singularity, where nothing resembles what goes on in the Newtonian world. Time (coordinate time!) stops there and Schwarzschild's sphere seems impenetrable. But, as all good (and not-so-good) authors observed, who cares, since that sphere remains a virtual one, hidden under the surface of the star—since that strange place can never manifest itself.

Moreover, the impenetrable nature of Schwarzschild's singularity—the fact that at that place time seems to stop—*appeared* to be a provable proposition. But there were mathematical proofs that did not correctly account for the physics, and this was not always easy to see. Only much later was it realized that the problem stemmed from a *poor* choice of coordinates. Schwarzschild had chanced upon coordinates that did not cover the *whole* space, in particular the region of space hidden behind his sphere. As soon as trajectories approached Schwarzschild's singularity, they got stuck, so to speak. This was then taken as bona fide evidence that all trajectories ended up or died at the singularity, where time (coordinate time!) stopped; for if time stopped, how could anything still move forward? The trajectory appeared to perpetually approach the magical sphere, as if to vanish there.

Zeno! The question here is precisely of the same type as the one arising in Zeno's paradox of Achilles and the tortoise. Achilles runs after a tortoise that has run off in front of him and seems unable to catch up with it, no matter how hard he tries. As Achilles travels the distance from his starting point to the starting point of the tortoise, the tortoise advances a certain distance, and while Achilles traverses this distance, the tortoise makes a further advance, and so on, ad infinitum. We have here what is known as a series, an infinite sum of numbers each of which is smaller than the preceding one but whose sum is finite, a fact that was not understood at the time. Achilles will always have some distance to run, even an infinitesimally small one. Zeno's calculation (which has to do with the fact that the sum of an infinite series of smaller and smaller numbers may have a finite limit) seems to imply that Achilles will never catch up with the tortoise. And yet he does! What happened is that the problem was incorrectly formulated. Just as Achilles does catch up with the tortoise, our particle enters the magical sphere . . . provided we give it enough (proper!) time. Time does not stop on the magical sphere, and our observers will soon disappear there.

DEAD ENDS IN SPACE-TIME

At the beginning of the 1920s, von Laue (then one of the leading specialists of the theory, whom we have already met) studied the trajectories of light particles—what we would now call photons—in Schwarzschild's field. Inspired by previous authors, he produced a decent and representative piece of work. His diagram (figure 12.2) represents the set of possible trajectories of the solution projected onto a plane passing through the center of the star. He drew all the trajectories approaching the central mass and, as can be readily seen, *none* of the light particles reached the center, $r = 0$; they all stopped at the surface of Schwarzschild's sphere. This was, therefore, a magic sphere on which time was supposed to stop and inside which nothing apparently happened.

I have also included a second figure from the same period (but we can still find this sort of representation at the beginning of the 1960s) that was in the thesis of a Belgian student, Carlo de Jans. In his diagram, the particles are material particles, but the same phenomenon takes place: all trajectories stop at the surface of Schwarzschild's sphere. But de Jans took his study a bit farther

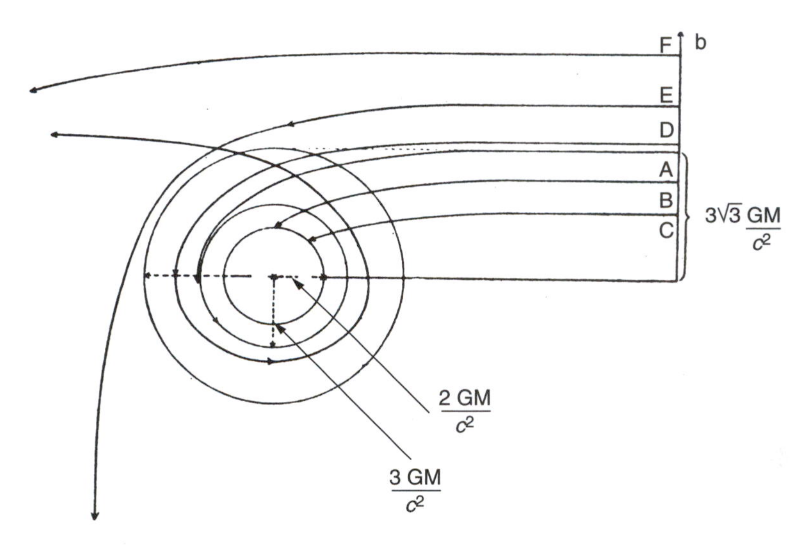

Figure 12.2. Light trajectories in Schwarzschild's space. The diagram shows these trajectories on a plane passing through the center. The small empty circle at the center of the diagram represents Schwarzschild's sphere. (From von Laue, 1921, p. 226)

than his teachers and colleagues and in his thesis developed a very thorough—and, incidentally, excellent—analysis of the subject; not a very appealing subject, we have to admit, but one whose study was necessary. Now, de Jans employed a parameter better suited to the situation than coordinate time. In the case of a particle spiraling toward the center of the solution, he used a natural parameter: the polar angle from the center of the revolving particle. And his calculation shows (to whoever examines it carefully) that the trajectory penetrates the impenetrable singularity without any resistance. But there is no one so deaf as he who will not hear, and this budding relativist reverted to the interpretation favored by his colleagues and his master, Théophile de Donder; he reverted to the parameterization in coordinate time t, without realizing that he had obtained an unusual result, to say the least, that contradicted his conclusions. But he was not the only one to go around in circles. Oh Zeno, how cruel you are!

Let us also briefly mention the work (once again unconscious!) of a little-known German astrophysicist, E. Rabe, who, soon after

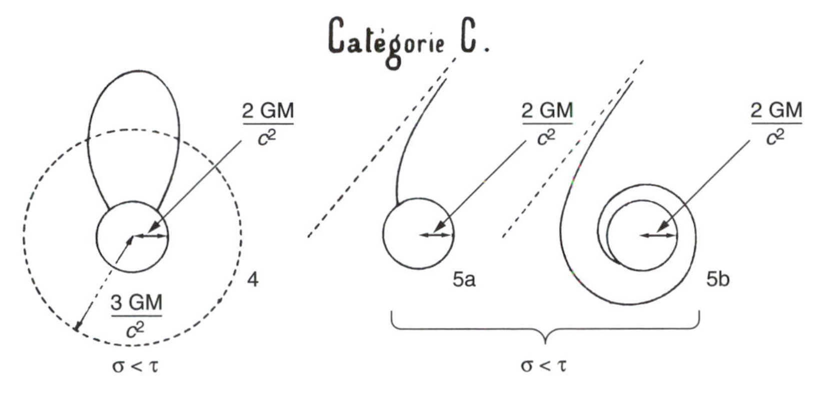

Figure 12.3. Trajectories of material particles in Schwarzschild's space. These are examples of trajectories drawn on a plane passing through the center. The small empty circle represents the impenetrable Schwarzschild's sphere. (From de Jans, 1923, plate 3)

World War II, studied the (light or material) particles directly falling, in free fall, toward the center of the sphere: the question of falling bodies in general relativity.[3] He selected, quite naturally but somewhat by chance, a good parameterization, and the trajectory penetrated Schwarzschild's sphere, which he took, of course, for a singularity. However, he had no merit, for he did not notice anything, given that in this very simple and particular case, the equation of a falling body in a Schwarzschild space has precisely the same (simple) form as the corresponding Newtonian equation. Once again, just like his colleagues, he reverted to the wrong parameterization in order to prevent the particles from reaching the region proscribed by his peers.

Let me add that it is not with malicious intentions (without which there is no good literature) that I chose these tortuous and hesitant examples, but rather because they reflect the relativists' indecisiveness and the manner in which they temporarily settled these issues. Naturally, I could have presented the correct solution right away, as texts appropriately do today, and I will do so next. But it is useful to wander off course, as the best minds did, in order to better understand the problem and learn that we are not the only ones to be puzzled. Doing so makes for a relaxing diversion from the anguish of space-time. At any rate, the problem of falling bodies in general relativity, that is, "Galileo's problem," if I may call it so, for that is precisely the question here, was not really solved until the 1960s.

And yet, Georges Lemaître, the cosmologist abbot, showed very neatly that Schwarzschild's sphere was not a singularity in the mathematical sense, but no one paid attention. He didn't help matters by publishing his 1933 article in French and in a Belgian journal! German was still the dominant language in physics at the time, although soon to be replaced by English. Einstein had just left Berlin for Princeton, and Nazism would soon pervade Germany and Europe. But let us go back to space. The problem is not as simple as one would have expected; to fully understand it, it was not enough to use the right parameterization or to study the trajectories one by one, each with its proper time. It was also necessary to correlate all those proper times and all those trajectories against each other, to integrate all those paths, in a global fashion, into a single space-time. The physicists were facing a structural problem, that of the topological structure of space-time. A new global system of coordinates was required, but this would not be realized until 1960 and would result in a very strange space-time. Behind this seemingly minor problem of Schwarzschild's singularity, it was the global architecture of the solution that was at stake—as well as the future of general relativity.

In hindsight, it is clear that the fundamental mistake in analyzing the question has stemmed from a badly chosen parameterization—coordinate time instead of proper time—but also, and above all, from a lack of understanding of the very essence of the problem from both a physical and a philosophical angle. We must make sure that the *entire* history of *each* particle in motion through space has unfolded; that we know each complete trajectory, each whole life, so to speak; and also that we find a way to map all those paths together by creating an atlas, a collection of maps of all possible and imaginable trajectories. Each trajectory should span the entire range, from $\tau = -\infty$ to $\tau = +\infty$, of its proper time. In other words, we must know its history from the beginning to the end of its proper time. By collecting all those trajectories, all those maps, in an atlas, we obtain the so-called maximal extension of space: the entire accessible space-time.

All this is not, as the reader might too quickly suspect, the wooly ramblings of a philosopher in want of a subject. The principle of maximal extension is an extremely important one, used today to analyze topological questions in general relativity. Words such as "map" and "atlas" are defined mathematically and widely used in relativistic works, in which we also find an obvious and natural idea: space is the set of all accessible places (in the past or in the future). In order to discover space, we must travel in it. Philosophy is at work here. Each particle must necessarily have both a future and a past, and its path must be traced out from the beginning of the universe—big bang or infinitely distant past—to its end—big crunch or infinitely distant future. Isn't that what the principle of conservation of matter requires? The principle of geodesic extension must take into account the crucial question of singularities in general relativity. Better still, it enables us to put things in perspective and to specify what a singularity is (and what isn't). Thus, we will follow the path of each particle from the beginning to the end of their history; through their proper time, τ, from $\tau = -\infty$ to $\tau = +\infty$; from the infinite past to the infinite future, unless the particle comes from (or falls into) a true singularity—for we cannot expect a particle to travel through the big bang. The big bang, which, as we know (see chapter 15), represents the moment when the universe is reduced to a point and hence when space-time is really singular.

Big bang, big crunch:[4] these are real singularities (and sore points of the theory!). Allow me to fast-forward to the so-called *standard* cosmological model (more on this in chapter 15) with its *real* singularity: the initial cosmological singularity—the *big bang*. In that amazing place, space-time is truly, absolutely, and definitely singular. This is no small problem because there the theory becomes meaningless and space-time is not even defined. In that strange place (but is it even a place?), nothing can move, nothing can be, nothing can reach it—and it is the same thing at any *real* singularity. There is another type of singularity, the so-called *apparent* singularities (such as Schwarzschild's), through which things can travel. That is the simplest way to distinguish between real and apparent singularities.[5] The definition of maximal extension of a trajectory in space-time will have to conform to the fact that all geodesics must be allowed to reach their maximal extension, that is, their proper time (τ) must run from $\tau = -\infty$ to $\tau = +\infty$, unless they begin or end at a *real* singularity. This is in fact a definition of space-time itself: the set of all possible trajectories, and it implies that space-time is not at all known in advance but will only be defined at the end of a theoretical journey; it is the future of the theory, its conclusion.

Inside Schwarzschild's *pseudo*-singularity, at the center of his solution (at $r = 0$), there is also a (real) singularity. It is the heart of the black hole, a catastrophe of the same order as the one Hadamard anticipated, the place where the unfortunate particles that would happen to cross "Schwarzschild's horizon" would crash. *Horizon* is the term used today to designate it, in the context of the new interpretation of the solution; it is the apparent singularity, Schwarzschild's pseudo-singularity, the magic sphere. *Horizon* is a term that justifies the analogy I presented above between the evolution of the representation of the Earth and that of Schwarzschild's solution. We shall return to the subject in the next chapter, that is, to the description of the solution in terms of the topology of space-time. Until then, let us examine another, more physical and absolutely essential fact: the effect of gravitation on light. This will allow us to consider the question from a more realistic perspective and to appreciate the need for all those difficult and somewhat abstruse works.

DARK BODIES: THE ANCESTORS OF BLACK HOLES

At this point, we must go back to the beginning of general relativity, in particular to the deflection of light by the Sun's or some other star's gravitational field. As we saw in chapter 8, since

1907, Einstein had been aware that in his still nascent theory light would have to be deflected by gravitation. I mentioned his early calculations of 1911, which were developed in the Newtonian context, and the final result he obtained later, in 1915, in the framework of general relativity, the formula that was to be verified during a solar eclipse. This formula, which gives the deflection of light in the gravitational field of a mass M at a distance *d* of its center, is written $\delta = 4GM/dc^2$. The relativistic deflection (twice the Newtonian value but of like structure) is therefore proportional to the mass M of the star and inversely proportional to its distance *d* to the center. But this formula is only an approximation, obtained assuming that δ is small (1′70″)—which is correct, if the measurements are taken near the Sun—and that it is enough to compute the first-order term, the higher-order terms (in δ^2) being considered negligible. In the context of the calculation of trajectories in Schwarzschild's space, all light geodesics were very accurately determined in the 1920s and justified the use of the above approximation, which would not be challenged for the next fifty years.

But if this analysis is adequate in a weak gravitational field (such as on the Sun's surface), is it valid in general? The question naturally arises of the effect of gravitation on light in the case of strong fields. What happens to the above formula (and to the light beam it describes) as we approach the field's source, which we may assume is becoming more compact, that is, as $2GM/dc^2$ increases? This is precisely the question von Laue evaded in his textbook: What happens on the pseudo-singularity we shall from now on call the *horizon*? What really happens to the trajectories of light rays if we "parameterize" them correctly? It is easy, the reader would object, to ask that question today, in hindsight. In order to better grasp what happened in that case, we must return to the eighteenth century, when in the context of Newton's theory, some natural philosophers understood certain aspects of the question better than did the relativists.

The story of dark bodies is interesting not merely for historical reasons but also for pedagogical ones, for it will help us understand one of the fundamental aspects of the phenomenon within the framework of Newton's theory through more accessible and familiar concepts. Distances, velocities, accelerations, all concepts that disappeared in Einstein's theory, will prove very useful. Very useful and very dangerous, for they will enable us to think in Newtonian terms, and that is precisely what I strongly denounced above. If we deliberately think of the effect of gravitation on light in those terms and take a look at dark bodies, it is expressly for the purpose of (figuratively) throwing some light on their cousins, black holes; however, we shall very clearly establish the limits of this analogy. Although we have done so throughout this book, it will be helpful to recall the most important points.

In his *Principia*, Newton developed, in fact, *two* theories—gravitation and his corpuscular theory of light—precisely because he treated light as particles subject to a dynamics of *refringent* forces, that is, in the same way he treated material bodies that are subject to gravitational forces. To sum up, Newton treated light particles as material particles in a sort of ballistics of light.

We must not forget that light particles were then subject, just as any particle, to Galilean kinematics and not to the kinematics of special relativity! As we pointed out in chapter 1, light was not supposed to have a constant speed and could be accelerated or slowed down. Moreover, light particles had mass, and there were concerns regarding the loss of mass the Sun suffered every day. And if light was a particle like any other, why shouldn't it be subject to *universal* gravitation? There was no reason why it shouldn't. It was a typical Newtonian assumption, one which Newton himself had entertained.

In mid-eighteenth-century England, John Michell, a minister, excellent astronomer, physicist, and natural philosopher, defended Newton's theory in earnest. He had studied at Cambridge, where one hundred years earlier Newton himself had studied and taught. He was the first, after Newton, to seriously consider the effect of gravitation on light. Emitted on the surface of a star with a constant emission speed c_0, a light particle was slowed down by the gravitational forces acting on it. Michell used geometrical methods to carry out his calculations because algebra was not yet part of an English philosopher's mathematical toolkit.

He restricted himself to the case of a light particle emitted along a radius by a star (not necessarily the Sun) with an emission speed c_0. Assuming the mass of the star to be known, say, M, Michell calculated the speed of the light particle when it was at a distance *d* of the center of the star. It was an extremely simple calculation for a specialist of the Newtonian theory of gravitation such as Michell. The light particle was slowed down by the star's gravitational field, just like a stone that had been thrown into the air; a very, very light stone, but one which was thrown with an extremely high initial (emission) speed. The result was not surprising and depended on the mass M of the star, its radius, and, of course, the emission speed.

In his calculations, Michell used the speed gathered by "a body falling from an infinite height"[6] toward the Sun, which was precisely the speed of a body ejected from the star so that it would never fall back to it. It did not take him long to notice that this "diminution of the velocity,"[7] as he called it, might very well become zero.

Thus, light particles emitted with speed c_0 by a giant star would be gradually slowed down until their speed became zero at a certain distance from their source; beyond this distance, the star would be invisible. Michell was well aware of all this, and he soon used such invisible stars to explain the possible existence of gravitational fields whose source was invisible: dark matter before its time. Dark bodies—as Laplace would call these hypothetical stars—were born. Laplace would also promote them in his splendid popularization *Exposition du système du monde*,[8] which was published during the French revolution, but without ever mentioning the name of their inventor.

In the eighteenth century, three possible effects of gravitation on light were envisaged. In 1772, Michell foresaw the gravitational slowing down of the light emitted by a star. In 1791, William Her-

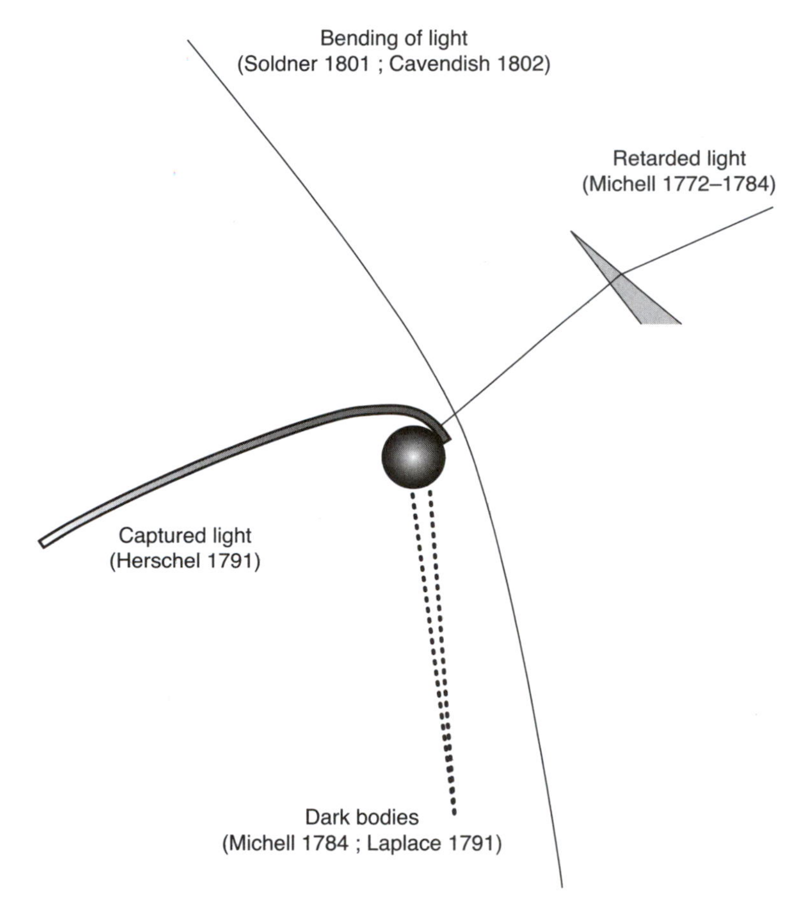

Figure 12.4. The effect of gravitation on light in a Newtonian context.

schel believed that light might be captured by a planetary nebula, and in 1801, Soldner calculated the deflection of light by a heavy body.

Before leaving the eighteenth-century perspective, we must note that dark bodies were not black holes, primarily, because the physical concepts upon which these strange objects depend do not come from the same theory. Michell spoke of the speed of a particle becoming zero; this is a heresy in general relativity where, as special relativity requires, the speed of light is always

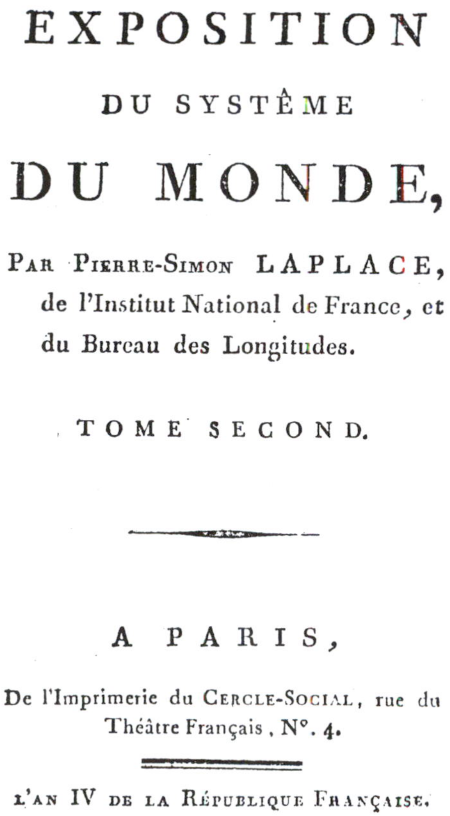

EXPOSITION

DU SYSTÊME

DU MONDE,

PAR PIERRE-SIMON LAPLACE,
de l'Institut National de France, et
du Bureau des Longitudes.

TOME SECOND.

A PARIS,

De l'Imprimerie du CERCLE-SOCIAL, rue du
Théâtre Français, N°. 4.

L'AN IV DE LA RÉPUBLIQUE FRANÇAISE.

Figure 12.5. Dark bodies in Laplace's *Exposition du système du monde*, 1796, volume 2, title page and p. 305. (Photograph Observatoire de Paris)

and everywhere constant and equal to c, even if light is subject to the gravitational field. Moreover, as we shall see in the next chapter, all particles, light particles or material ones, disappear at Schwarzschild's horizon without their speed becoming zero; far from it.

But there are real similarities between dark bodies and black holes. Even if the theories are different—and they are extremely so!—the physical phenomenon supporting these theoretical constructions is the same in both cases: the effect of gravitation on light. Light is affected by gravitation, as Einstein showed and

Tous ces corps devenus invisibles, sont à la même place où ils ont été observés, puisqu'ils n'en ont point changé, durant leur apparition ; il existe donc dans les espaces célestes, des corps obscurs aussi considérables, et peut être en aussi grand nombre, que les étoiles. Un astre lumineux de même densité que la terre, et dont le diamètre serait deux cents cinquante fois plus grand que celui du soleil, ne laisserait en vertu de son attraction, parvenir aucun de ses rayons jusqu'à nous ; il est donc possible que les plus grands corps lumineux de l'univers, soient par cela même, invisibles. Une étoile qui, sans être de cette grandeur, surpasserait considérablement le soleil ; affaiblirait sensiblement la vîtesse de la lumière, et augmenterait ainsi l'étendue de son aberration. Cette différence dans l'aber-

Figure 12.5. (*Continued*)

which was later verified in 1919 at Sobral and elsewhere (see chapter 8). Also, the deflection of light was implied, qualitatively, by Newton's theory, a fact that Eddington had perfectly understood. As for another similarity, both involve a star whose gravitational field is strong enough to trap light particles, although in the case of dark bodies this trap is not as hermetic as the one due to black holes: light may escape from a dark body but not from a black hole.

Eighteenth-century works not only provide us with the images to think about dark bodies and black holes, but they also make us wonder: how is it possible that these strange objects could have been imagined back then while relativists would balk for fifty years before envisaging a phenomenon that was, after all, quite similar?

SCHWARZSCHILD'S SOLUTION

The metric of Schwarzschild's solution represents the gravitational field created by a spherical distribution of mass, such as a star, the Sun, or the Earth:

$$ds^2 = \left(1 - \frac{2GM}{rc^2}\right)c^2dt^2 - \frac{dr^2}{1 - \frac{2GM}{rc^2}} - r^2(d\theta^2 + \sin^2\theta d\phi^2)$$

As we have seen in chapter 7, solutions in general relativity are generally expressed through the so-called *line element of space-time*. This mathematical object is the infinitesimal proper time, *ds*, between two points of space-time and it is used to calculate proper time.

This solution, or ds^2, as it is called in technical jargon, takes a simple form thanks to the use of polar coordinates, which are the usual coordinates in astronomy and rational mechanics for the study of this kind of problem; r represents the distance to the center of the star, t is coordinate time, while θ and ϕ are angular coordinates similar to latitude and longitude.

In fact, we can ignore the angular variables, which do not cause any problem, and focus on the time and radial coordinates. This amounts to studying the gravitational field from the center of the star to infinity and for all time. To this end, we must calculate the proper time between two points of space-time located at different altitudes and at different times.

But do these coordinates have an intrinsic physical meaning? Do they have the same interpretation as in Newtonian theory or even in special relativity? This was one of the most difficult questions to understand and accept. Einstein himself was reluctant to accept his own clear stipulation and what is now the creed of present-day relativists: coordinates do not have any a priori physical meaning.

Things would be easy if Schwarzschild's solution had a familiar structure, but this, alas, is not the case. The problem comes from the term $1 - 2GM/rc^2$ expressing gravitation, which appears both in the numerator of the gravitational potential associated with time (the coefficient of the dt^2 term) as well as in the denominator of the gravitational potential associated with the radius (the coefficient of dr^2 term).

First notice that if M is zero, we simply obtain the line element of special relativity (in polar coordinates, of course). The same is true when r approaches $+\infty$: the term M/r then becomes arbitrarily small and the coefficients of time and space—that is, the gravitational potentials—are (at infinity) equal to 1. This means that space is flat, that there is no gravitation, only inertia, which is reassuring, for if we are at infinity or if the central mass is zero, it is normal and desirable that the resulting space-time should be flat.

But if we take a look at the situation at the center ($r = 0$) or near the center, when $r = 2GM/c^2$, the gravitational potentials take a disturbing form. When $r = 2GM/c^2$, the coefficient of the (infinitesimal) radial variable dr^2 becomes infinity and that of dt^2, the time variable, becomes zero and then changes sign for r smaller than $2GM/c^2$. In other words, from a mathematical point of view, the *gravitational potentials* are singular: one of them because it becomes infinity, and the other because it changes sign; likewise at the central singularity, at $r = 0$, where one of the potentials becomes infinity and the other zero. However, this was wrongly considered not to be a serious problem, because some singularity was to be expected at that place—just as in Newton's theory, where the gravitational potential also contains the factor $1/r$ and is therefore singular at $r = 0$—but has today become a major concern: in general relativity space-time must be *everywhere* defined, so what is this singular place where space-time vanishes? Where and how would space-time cease to exist?

It is an entirely different story at $r = 2GM/c^2$, at the so-called Schwarzschild's singularity. Fifty years later, it would be around this singularity that the solution would begin to be built: the interpretation of the solution in terms of a black hole. But, for the time being, relativists could only speculate as to the meaning of this weird object, and the most bizarre interpretations of this somewhat fantastic spot were proposed. By interpreting it in terms of coordinates, they tried at all costs to show that it was an

Figure 12.6. Karl Schwarzschild. (© AIP Emilio Segrè Visual Archives)

impassable barrier. And it is easy to understand why: not only do the potentials become infinity there, but beyond this point (and certain coordinate systems will allow this), the signature of the line element of space-time, the all-too-famous ds^2, changes from (+ – – –) to (– + – –). This would mean that time becomes space and space becomes time, for, if coordinates have no physical meaning, a point on which Einstein rightly insisted, the only way to tell time apart from space is by their sign in the local definition of ds^2.

NOTES

1. The reader will find a more detailed analysis in the author's articles on the subject: Eisenstaedt, 1982, 1987.

2. Nordmann, 1922, p. 155.

3. Cf. on this topic, Eisenstaedt, 1987, pp. 312–328.

4. This also applies to the big crunch: the only difference between these two singularities is that the first one represents the beginning of the universe, and the second one its end. But in fact, these true singularities represent limits of general relativity, and theoretical physicists cannot ignore this situation. In the vicinity of these singularities the density of matter becomes phenomenal, and quantum effects must then be taken into account; at the singularity itself, the density is infinity, which is physically unacceptable. It is then necessary to appeal to a unified theory of physical forces, including gravitation in order to study what goes on at that place from a quantum perspective. Cf. chapter 16.

5. In fact, in general relativity there is no precise mathematical definition of "singularity," that is, of a notion of singularity general enough to apply to all problems of the theory. General relativity, just as practically every other physical theory, is conceptually incomplete. It is therefore always a "work in progress." But this is not restoration work—as in the case of our cathedrals, whose original state we piously try to preserve—but of fundamental work: we never finish working on a physical (or mathematical) theory; it can always be improved, and that is precisely the task of our specialists.

6. Michell, 1784, p. 42. Cf. Eisenstaedt, 1991.

7. Michell, 1784, p. 50. We can easily express Michell's calculations in algebraic terms, precisely as Laplace did a few years later (Cf. Laplace, 1799).

The speed of light at a distance r from a mass M of radius r_0, with an emission velocity c_0 is equal to $c = \sqrt{c_0^2 - \frac{2GM}{r_0} + \frac{2GM}{r}}$.

Hence, the speed $c(r)$ of light can become zero if $c_0^2 < \frac{2GM}{r_0}$, that is, if $\frac{2GM}{r_0 c_0^2} > 1$. We find here, in a purely Newtonian calculation, Schwarzschild's radius (cf. box 5, chapter 7). In this context, we may say that a stone falling from infinity to $2GM/c^2$, that is, to a distance equal to the Schwarzschild's radius from a star or a planet, would arrive there with a speed equal to that of light. This Newtonian calculation is obviously enlightening for the relativistic interpretation, for it suggests a connection between Schwarzschild's radius and the behavior of light.

8. Laplace, 1796, pp. 304–306; cf., in this respect, Eisenstaedt, 1991, 1997.

CHAPTER THIRTEEN

Paths in Schwarzschild's Space-Time

IN ORDER TO better understand the concept of a black hole and grasp both its necessity and its strangeness, let us go back to the structure of Schwarzschild's sphere as it was understood from the 1920s to the 1960s and represented in von Laue's and de Jans's diagrams (figures 12.2 and 12.3). It was a magical sphere, a sort of impenetrable ball on which light and material particles accumulated.

The trajectories, as shown in those figures, have two possible limits or "infinity." The spatial infinity of the familiar (if we may call it so) flat, Minkowskian space, and the internal limit that is this magical sphere, where the trajectories seem to vanish. They *seem* to vanish, because, contrary to what was then believed, this magical sphere is not singular and nothing really terrible takes place there; the particles simply travel on—well, almost, as we shall see later. In fact, particles may in fact go through the surface and crash at $r = 0$, the center of the solution, a truly singular point strongly resembling what is now known as the big bang.

Let us recall our creed: all trajectories must be *extended* through their entire parameterization, from the infinite past to the infinite future. But it may happen that a trajectory should come across a snag, a "real" singularity such as the big bang. In that case, we will accept that it should not begin at the infinite past or reach the infinite future, but that its journey should end at this (real) singularity. For all (light or material) particles, all particles in the universe, must have a past and a future; they must become "something," if only by virtue of the principle of conservation of matter and energy. They must "become, be, and have been," unless they end up at some singularity where nothing makes sense any more. If we don't know how to define what a real singularity is—as I observed in the previous chapter—at least we know what it *isn't*. In short, we know what we don't want, and Schwarzschild's sphere is not singular; it is an apparent singularity (and for this reason it is now called "horizon"), whereas the point (the line) $r = 0$ of space-time, the origin of the coordinate system, is indeed singular. The reason for this distinction is, basically, that trajectories may go through the "horizon" but they stop at $r = 0$ (where else could they go?), but also because all the intrinsic magnitudes that relativists could imagine are, from a mathematical point of view, regular on Schwarzschild's sphere but singular at the origin.

Thus, there is nothing singular on Schwarzschild's sphere, and so there is no reason why our particles should suddenly finish their life there, as de Jans had in fact shown (without seeing it). Coordinate time is not physical time but, rather, a misleading parameter (I sufficiently stressed this point, Zeno!), and with a proper parameterization, the trajectory may be extended beyond the horizon and into the magical sphere. And if particles can enter the sphere, it means that there is space on the other side, that something goes on in there, but what, exactly?

The particles of the collapsing star do not die on the magical Schwarzschild's sphere, on the horizon; they merely seem to disappear on this surface only to reappear on the other side and finally end their life at the center, at the real singularity (figure 13.1). Weird! But there is something still stranger, for the particle seems to make the last part of its trip by going backward in coordinate time. In fact, as the particle falls onto the magical sphere, it simultaneously travels from the origin toward this sphere, where it meets . . . itself. Weird! Weird! This is science fiction stuff. And the attentive reader may have plenty of questions. . . . But let us keep things in perspective and not forget that what we are describing here is the way these questions appeared in the years 1930–1950, when the problems raised by Schwarzschild's solution were discovered, that is, before the invention and the acceptance of the black hole as the—strange, we must admit—solution to these difficult questions.

Let us resume our argument at the moment where the particle meets itself as if it would meet its own image reflected in a mirror. As the particle traverses the horizon (as we have seen in chapter 12), the time coordinate t and the radius coordinate r ex-

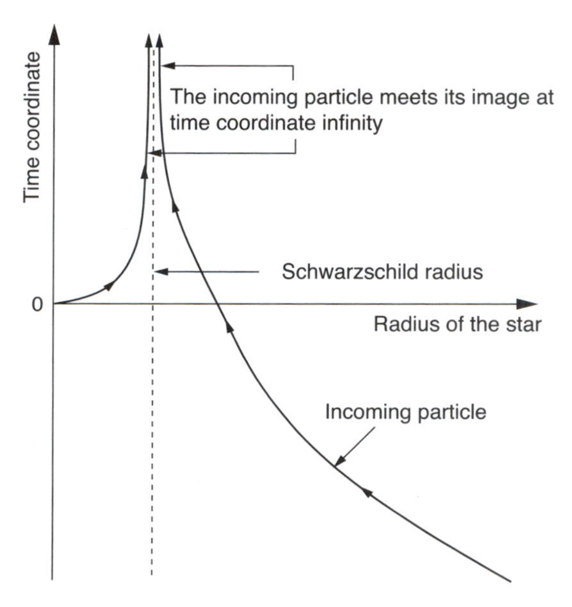

Figure 13.1. A trajectory in Schwarzschild's space. The incoming particle meets its own image on the magical sphere when the coordinate time *t* becomes infinity. Likewise for the outgoing particle, not represented here. (From H. P. Robertson and T. W. Noonan, 1968, p. 249)

change roles, because their respective signs in the definition of the line element of space-time change. Fortunately, the description based on the proper time of the particle in free fall poses no problem, and it accounts for the trip of the particle through the horizon and up to the center without any paradox. However, once the (light or material) particle has crossed the horizon, it becomes invisible to any observer outside the sphere. Isn't this a simpler and more convincing explanation? Coordinate time *t* turns out to be completely inadequate to handle the problem.

QUESTIONS FROM A COSMOLOGIST

If we had only listened to Einstein! Alas, even Einstein, who tirelessly repeated the same (and how right!) lesson ("coordinates have no physical meaning"), did not see how to apply it here. On this issue, he took a conservative position. Lost in this labyrinth, even he could not make head or tails of it. He did not understand and did not accept Robertson's work. Howard Robertson was a cosmologist whose office was next to Einstein's and who was trying to understand how all these trajectories could be put together.

And yet, Robertson's description was not complete, was not acceptable, since after he traced out *all* possible trajectories of the solution, he not only found particles "going into" the sphere, but also particles "coming out" from it which did not belong to the same diagram of space-time. And, conversely, in a diagram where the trajectories of outgoing particles were represented, those of the incoming ones had no place in it. Thus, he needed two diagrams, two maps, to represent the set of all possible trajectories in Schwarzschild's space.

But how, then, can one put together, how can one connect, those two maps in a consistent way to form a single "atlas"? One cannot do as one wishes with a solution of a theory, be it Schwarzschild's or some other one. A solution of a theory has a life of its own, and it will take you here and there without your being able to do anything about it. And if, as is the case here, the solution allows certain trajectories, you do not have the right to ignore them. It would take reasons we don't have or no longer have to do so, such as claiming that Schwarzschild's sphere would be singular, which is not true, as Abbot Lemaître had just shown. In short, there are particles coming out of the sphere but backwards in time, so to speak, rushing to meet their image that converges toward the horizon. What do we do, what are we to think when we realize that, even trapped in the black hole, they can still escape? That is exactly what happens. But how can it be possible, and where do they go?

All these questions will have to be reconsidered, maps and atlases will have to be redrawn, as when it was discovered that the Earth was round, in Parmenides's, Plato's, and Aristotle's time. Let us come back for a moment to the question of the representation of the Earth, for it can help us to picture that of the space-time around a star.

Most representations of the Earth are based on projections onto other surfaces, surfaces that we could unroll or unfold such as a cylinder or a cone and obtain a plane. A piece of paper (and some glue) is all we need to construct one of these surfaces. To obtain a plane, we would simply need to cut the cylinder (along its generatrix), and the same is true of the cone. A sphere, on the other hand, is not an "unrollable" surface, for it is impossible to unroll it into a plane without tearing it up. More important still, there is a fundamental topological difference between a plane and a sphere. A sphere is not topologically equivalent to a plane (figure 13.2); it cannot be transformed into a plane by a continuous distortion. (But a sphere *is* topologically equivalent to a cucumber or an onion.) A sphere is not homeomorphic or topologically equivalent to a plane, if only because it does not have points at infinity. And this is why we *need* more than one sheet of paper to represent the Earth in full, more than one map to represent all possible itineraries. There is necessarily a discontinuity somewhere—at the equator, the poles, or a parallel that is conveniently located on an ocean. This drawback can be resolved by resorting to an atlas, which enables us to glue the maps together, or by additional maps that illustrate the loss of continuity. We need at least two plane maps to represent, via projections, the Earth's two hemispheres, and we cannot do with only one, except if we accept that certain paths will be discontinuous.

Robertson's problem was not much different. He was forced to draw two maps, one for the incoming particles and another one for those going out, and in both maps there was a discontinuity at Schwarzschild's horizon. Not only was the continuity of the trajectories not reflected (even if he took it into account through a description in terms of proper time), but, what is more serious, he did not have an atlas that would allow him to understand how it was all put together. He did not even have the equivalent of a globe, which is a topologically correct representation of the Earth where all paths may be seen and understood.

Beyond the question of representation there remained that of the atlas showing the continuity of all paths in Schwarzschild's space. For Robertson, the only way out of this impasse was not

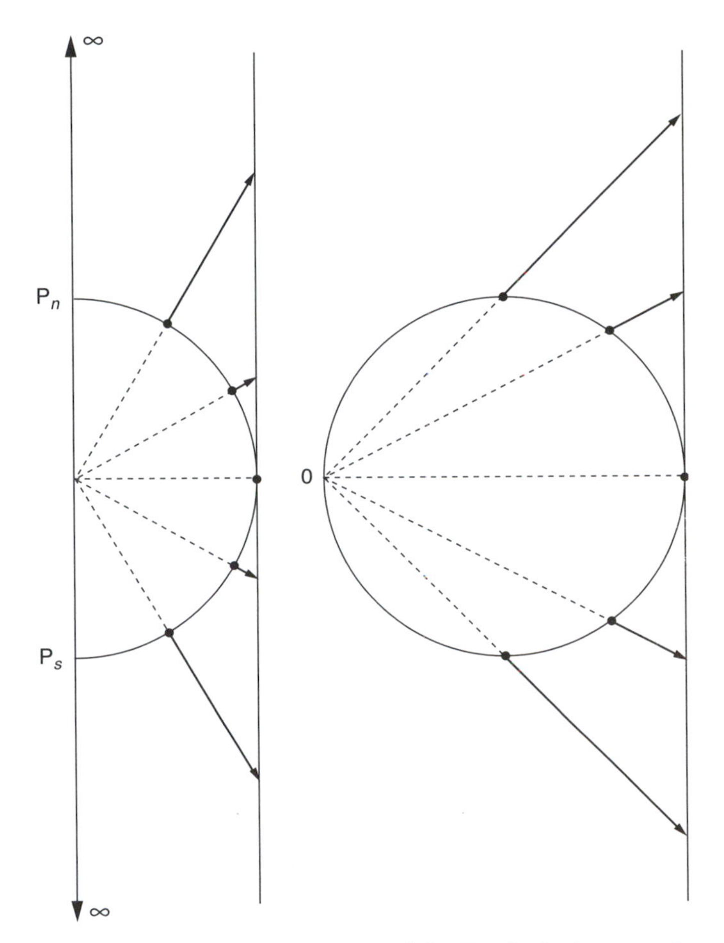

Figure 13.2. Representations of the Earth. Left: central cylindrical projection of the Earth. The projections of the poles (P_n and P_s) are at infinity. Right: the stereographic projection results in another distorted representation of the terrestrial sphere. The point of view, O, is sent to infinity. The images of certain continuous paths on the sphere are discontinuous on its representation. (Based on Joly, 1976, p. 64)

at all convincing: he clung to the fact that no physical object, star, material, or particle was known whose density was high enough for this singularity to occur; that such objects "do not exist"; that the physical radius of an object was always much larger than its gravitational one, be it a neutron or a star; in a nutshell, that

$1 - 2GM/c^2$ was in each case very close to 1 and hence far from the fateful zero. It was another way of saying that Schwarzschild's sphere, the horizon, was inaccessible or, in today's terms, that the existence of black holes was far from obvious. Therefore, all those questions were hypothetical and existed only in the mathematicians' dreams. Briefly: there was no problem. There was no problem at that time, in the 1930s, for, as far as dense stars were concerned, only the so-called white dwarfs were known and they did not even remotely cause any problem in that respect. For a body to have an effect, it would be necessary (and we shall come back to this below) to imagine some huge, immense star—an object we are not certain to have observed but may have come close to when we routinely observe pulsars, quasars, and other neutron stars. In the first part of the twentieth century, nobody wanted to believe it; the question of a dark star did not exist, and nobody wanted to remember that our natural philosopher, John Michell, had calculated its radius in 1784.

And so, relativists harped on this "it does not exist in nature"[1] for decades. We are tempted to ask: so what? For even if that were the case, even if their existence did not have any practical use, physicists would still have to answer the question of the consistency of Schwarzschild solution and therefore of Einstein's theory itself. This is no longer the case nowadays, for a number of exotic stars have been observed for which the ratio $2GM/c^2$ is not that far from R, their radius.

Two issues are now unavoidable: first, the structure of Schwarzschild's space-time and, second, the existence of a star with definitely exotic features. These two problems will dominate the work of relativists and astrophysicists, but only after the 1960s.

PENROSE'S DIAGRAM

Let us ignore for now the astrophysical difficulties and return to the structure of space-time by considering another representation, this one due to Roger Penrose. Figure 13.3 is a bird's eye

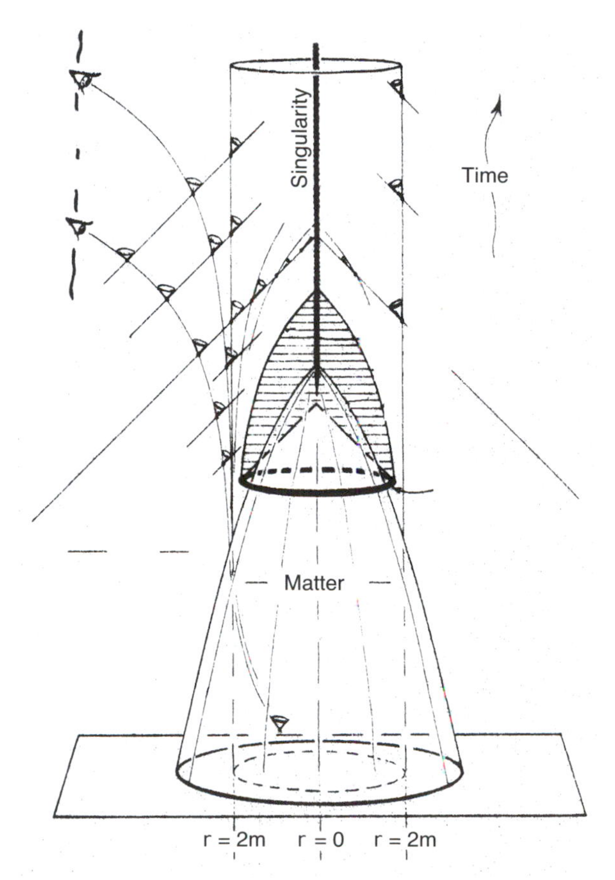

Figure 13.3. Another representation of the collapse of a star. (Based on Penrose, 1965)

view of a collapsing star. Two spatial coordinates are represented (a circle represents a sphere), and time flows from the bottom up in the diagram. The circle at the bottom represents the surface of the star, its radius, which is decreasing as the star collapses under its own weight. Then, the surface of the star goes through that cylinder representing the *magical* sphere, the horizon beyond which the star becomes invisible to an outside observer, until the star's implosion toward this dark line at the center of the diagram, which represents the real singularity at $r = 0$— where the star crashes and disappears. A few light rays are also represented, which must remain tangent to those small cones appearing throughout the picture. These cones, known as "light cones," are crucial, for they are made up of all the light rays passing through a given point of space-time (their vertex). All material trajectories must remain inside the cone, another way of saying that the speed of these is smaller than that of light. Thus, all light trajectories are tangent to some cone, and the surface of each cone consists of light trajectories.

Notice that the light cones are tangent to the cylinder representing Schwarzschild's horizon; but not all light rays, for it is clear that most of these fall into the black hole. Black hole? Yes, black hole, because by observing carefully one can see that no light ray or material particle can escape from the magical sphere, from the horizon. That's the reason why we cannot see what goes on inside . . . it is a black hole.

Schwarzschild's horizon is not singular in a strong sense, because space-time is defined on it and because it lets incoming particles penetrate it; it is a *unidirectional membrane*. But there is nevertheless something "magical" about this sphere, for it is composed of the light rays that determine its surface, and also because, just like a soap bubble, it is a perfect surface. It is also a mathematical surface, physically defined by the light rays that compose it, a bit like a straight line being a mathematical object but one whose physical reality manifests itself as a light ray (see chapter 6).

We have made considerable progress, for what we essentially need from a realistic point of view is all there: the description of the collapse of a star—its passage under the horizon, where it becomes invisible, followed by its final crash. Beyond the boundary of the visible universe, defined by the horizon, all that is left is the gravitational field of the vanished star. We shall return to this in the next chapter, in more detail.

And yet, this representation has one feature in common with Robertson's: it is not possible to show all trajectories in one diagram. In figure 13.3 we have represented only the incoming trajectories (the black hole), and there has to be a second diagram showing the outgoing particles—those leaving the real singularity—and that will represent what is called a "white hole."

These two maps describing the totality of the possible trajectories in Schwarzschild's space were connected in a final diagram that was proposed by M. D. Kruskal, a mathematical physicist, in 1960. Each particle in the universe must be accounted for, each of them must follow its fate in a space-time where all possible paths can be represented. That is called a *universe*. In a word, we will then finally know the shape of the space-time defined by a star.

KRUSKAL'S DIAGRAM

It was not until the end of the 1950s that things really moved forward and a solution was found. The article that marked and symbolized the renewal of general relativity was published by Kruskal in 1960. His paper, which at the time appeared difficult to a whole generation of relativists, today represents an acceptable interpretation of the theory. In my opinion, the acceptance of such an interpretation is due to the lack of a simpler, more convenient, and more familiar one, for Kruskal's is not an obvious and limpid solution to our problem. But it was imperative to find a way out and, as we are going to show, this is, despite its strangeness, a reasonable interpretation of a tough problem, the only consistent interpretation possible, although the debate over it is not closed.

What are we looking for? We want to define a space in which all possible and imaginable particles would have a past and a future ranging over their entire history—from the beginning to the end of their life span—until the end of time: an atlas. All particles, all the time. For it is definitely not acceptable that particles that have not reached a singularity should be prevented from moving along their path. And, in particular, we should not have to resort to *two* diagrams but, rather, a single one, which should be complete and readable and include all particles: those coming in as well as those going out.

Kruskal's diagram (figure 13.4) possesses all these properties, and it is the "clearest" available representation of the totality of

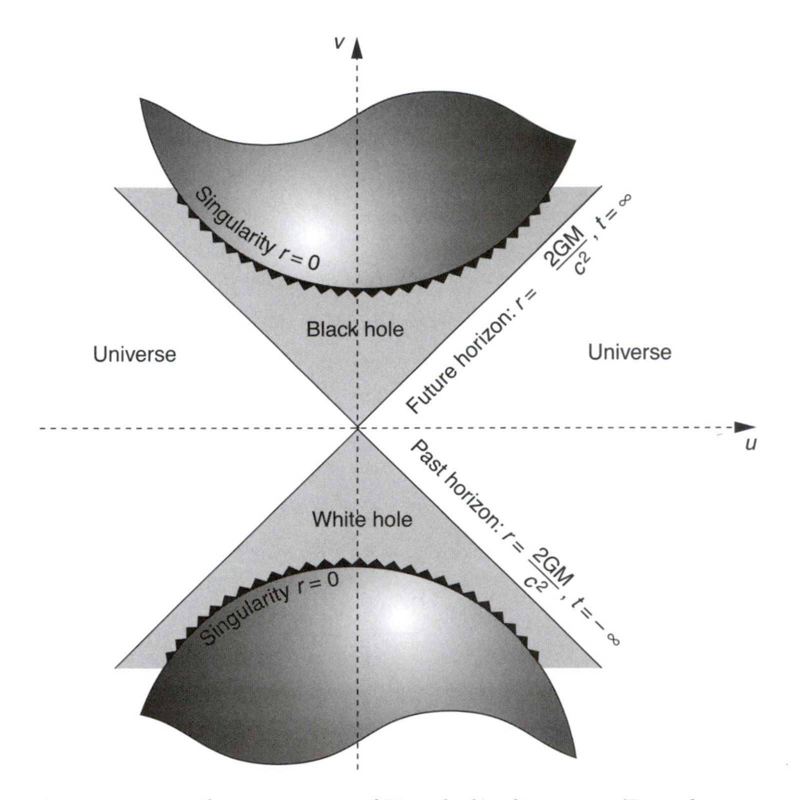

Figure 13.4. The structure of Kruskal's diagram. (Based on Kruskal, 1960)

Schwarzschild's space-time, the space created by a pointlike mass distribution at the origin.

Let us not try to read any physical meaning into the coordinates (u, v) of Kruskal's diagram; they are a reasonable combination of Schwarzschild's r and t coordinates. But it is at the price of abandoning any hope of making physical sense of coordinates (just as Einstein advocated) that we will be able to make sense of space-time itself. As in the preceding diagrams, the angular coordinates do not appear. The time arrow points upwards (in the diagram). The *two* diagonals represent the sphere $r = 2GM/c^2$ *split* into a future horizon and a past horizon, horizons on which the time coordinate t is infinity.

This representation of the geometry of Schwarzschild's empty space is symmetric with respect to the second diagonal (past horizon). The portion of the diagram in the third quadrant (lower left) is a mirror image of that in the first quadrant (upper right). This reflects the symmetry between outgoing particles: "white hole" (lower left), and incoming ones, "black hole" (upper right). Space-time is split in order to be able to show all trajectories and their maximal extensions. This point deserves to be emphasized: if theoreticians accept this splitting of space, it is not because they take a malicious pleasure in doing so but because they have to, in order to make sense of the trajectories of all particles in a single representation, that is, in the same space.

The dentate upper curve represents the real singularity, $r = 0$, into which the particles will eventually crash. This singular universe line has a mirror image in the lower half of the diagram: another dentate and singular line representing the white hole, the white hole out of which particles will come, while their counterparts will crash above, inside the black hole. This symmetry brings to mind Robertson's duplicate diagrams: on the upper right of figure 13.4 is represented the black hole of incoming trajectories and on the lower left the white hole of outgoing ones. Each trajectory has now both an entrance and an exit door, but above all, it has a past and a future. The splitting of space (represented by the symmetry black hole–white hole) is therefore the price to pay, the ransom for the consistency we have obtained. To be sure, we have not succeeded in attaching any physical meaning to this symmetry, and in particular to the region from which particles leave, that is, to the white hole. The dark gray areas are completely inaccessible; they are not part of the representation because they lie beyond the two essential singularities (one for each "hole") on which particles die or are born ($r = 0$). The light gray lower region represents the white hole, while the upper one represents the black hole. The diagonal labeled "future horizon" is the boundary between physical space, our physical universe, and the black hole. The past horizon is the boundary of the white hole. We shall see below that only a small part of this diagram is useful, that is, has a physical meaning.

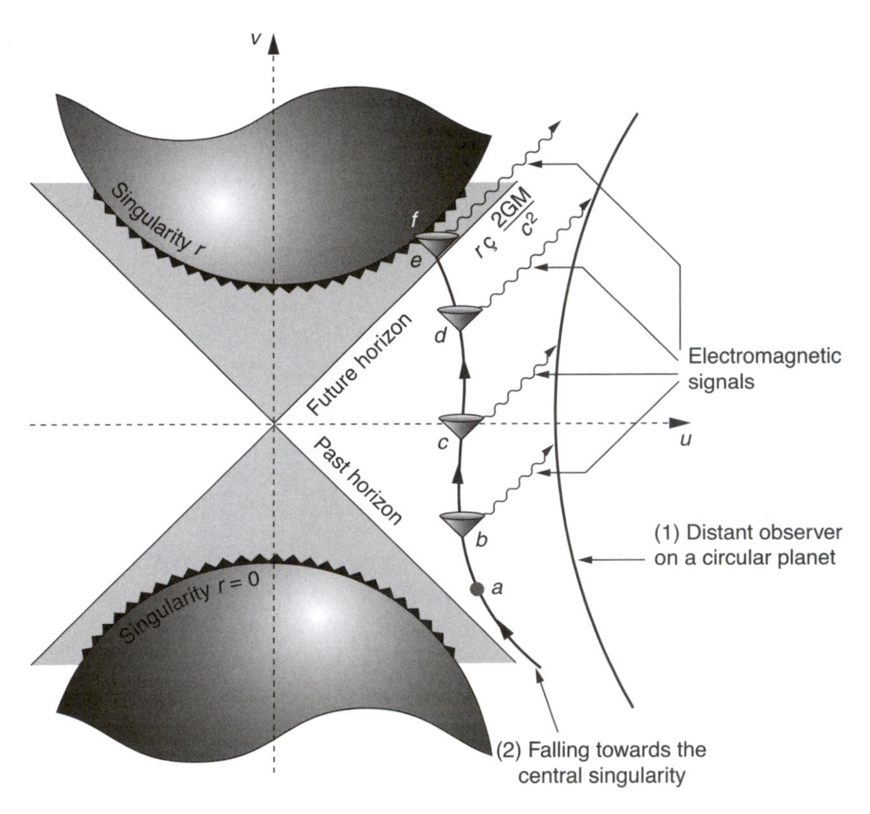

Figure 13.5. Trajectories in Kruskal's diagram.

In figure 13.5, we can see trajectories of (light or material) particles, labeled (1) and (2), calculated as a function of their proper time. They stretch over their entire particle life, their "maximal extension." They (simply!) represent falling bodies in general relativity. We are a long way from Galileo and the Pisa tower!

In this diagram, the cones are at 45-degree angles, parallel to the horizons: all electromagnetic signals, all light particles travel along diagonals, parallel to the cones and the horizons. Given that time flows upwards (in the diagram), with a little practice we can determine what information a given observer can receive or send and what he or she will ignore. An observer at the future horizon (which is simply our old Schwarzschild's singularity, the magical sphere—let's not forget that the two angular dimensions are missing here) will not be able to send any information to a distant observer on a circular orbit (on the right, in the diagram), as the cones situated on the trajectories indicate.

Let us now describe some trajectories of observers or particles in Kruskal's map (figure 13.5). The curve (1) represents the trajectory of a distant observer, a satellite, for instance, orbiting at a constant distance from the imploded star. This observer cannot see anything going on in the black hole but she can receive news from the white hole. Another observer, (2), is in free fall toward the horizon which he crosses at *e*. Once past *e*, he will not be able to send any information to the outside world. Then, he will travel inside the black hole and, after a finite interval of time (*e*–*f*) will end his life at the real singularity. The electromagnetic signals he sends all travel on the light cone, and only the signals originating at *a*, *b*, *c*, and *d* will reach observer (1) on her circular orbit far from the black hole. As for the signal sent from *e*, it will never be able to leave the black hole.

THE SPLITTING OF SPACE-TIME

Kruskal's representation implies the existence of two symmetric universes, or parts of a universe, one containing a black hole and another a white hole. It is possible to interpret these two universes as two parts, two pages of one and the same universe. Two global topologies have been proposed: one known as the "Einstein-Rosen bridge" identifies the two pages, while the other, the so-called "wormhole," connects the two pages by a sort of underground tunnel. The fact that it is possible to choose the topology, that is, the global structure or shape of space-time, stems from the form of Einstein's field equations; these are differential equations and therefore they only determine space-time locally. Figure 13.6 illustrates how the two parts of the universe may be connected and interpreted.

The diagrams in figure 13.6, showing two possible interpretations of Schwarzschild's space, are truly astonishing, especially when compared to its Newtonian counterpart, where a single curve is all that it takes to describe the same problem. How

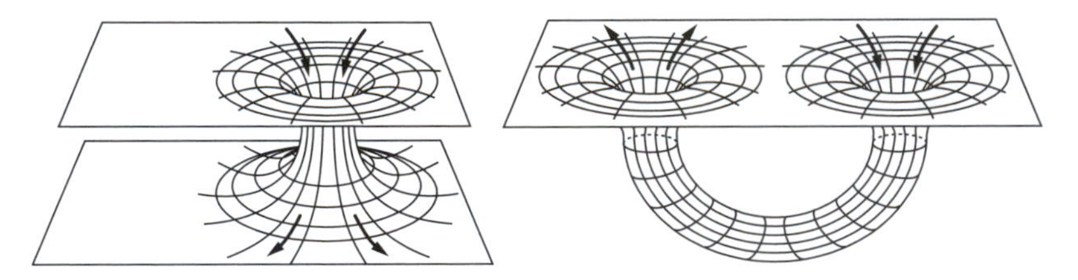

Figure 13.6. Two possible topologies: "Einstein-Rosen bridge" and "wormhole," or how to connect the two pages of Schwarzschild's space-time. (Based on C. W. Misner et al., 1973)

complicated things are in general relativity! We must understand and accept this new space and its trajectories—not an easy task for the young student or the attentive reader. We must also accept as necessary the analysis that led to that choice, especially the principle of geodesic extension, without which general relativity would be incomplete. Finally, we must be able to separate the wheat from the chaff: not everything is equally valuable in this diagram and not every component has a physical meaning. But what does the diagram tell us, then? In figure 13.7, Kruskal's representation of the collapse of a star, only the gray region admits of a convincing physical interpretation; for the rest of the diagram, in particular for the white hole, there is not yet any really convincing interpretation, either physical, astrophysical, or cosmological.

Let us set aside those hypothetical regions and those bizarre but necessary ramblings, and go back to more realistic problems, which are exotic enough. The gray zone in figure 13.7 corresponds to the classical region of space-time that extends into the black hole and has a convincing astrophysical interpretation (see chapter 14). This is the now classical representation of the collapse of a star resulting in the formation of a black hole, a prediction of general relativity favored by a very large majority of experts of the theory. This is also illustrated in figure 13.8, which is a physical representation of the collapse of a star and of various trajectories that are as realistic as possible.

Before this fateful year, 1960, when Kruskal published his incredible, immense—and even absurd, for some—interpretation

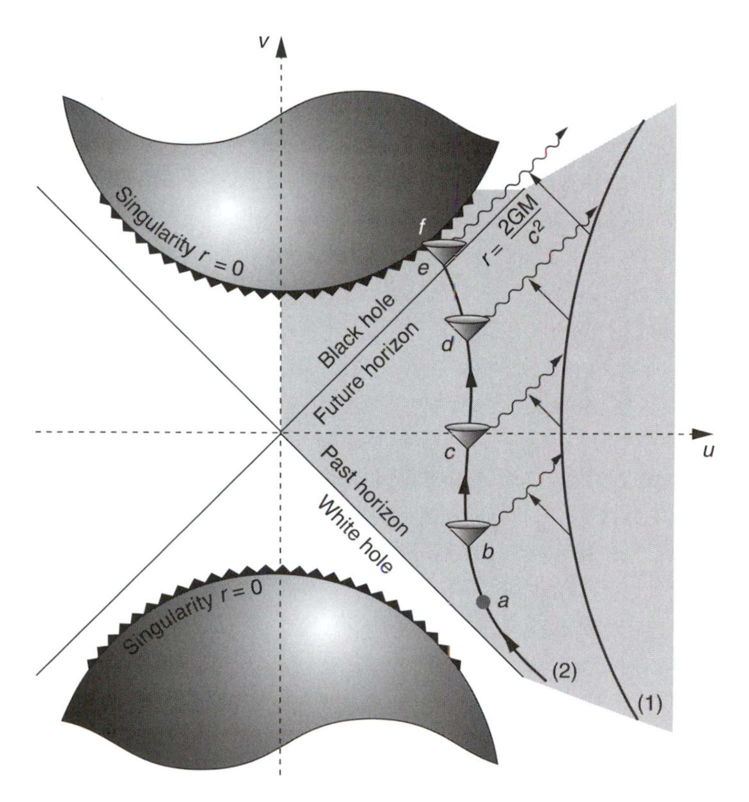

Figure 13.7. Kruskal's representation of the collapse of a star.

in *Physical Review*, very few relativists had imagined that things would come to that. Resting on their laurels, they were quite happy with a very gentle image of the theory and its three classical tests, which only slightly disrupted Newton's space-time and the serenity of those who still dominated the field in the universities: the specialists in rational mechanics and the astronomers—in a word, the "Newtonians." The field of gravitation was then almost completely dominated by Newton's theory, and Kruskal's interpretation came as a bombshell in the small and sleepy relativist village (Kruskal was a mathematician, and he probably did not realize the revolutionary impact his proposition would have). One day, someone will write the history of the 1960s, of that incredible period when the relativists, alarmed or delighted, but in any case divided, were confronted with that diagram, and

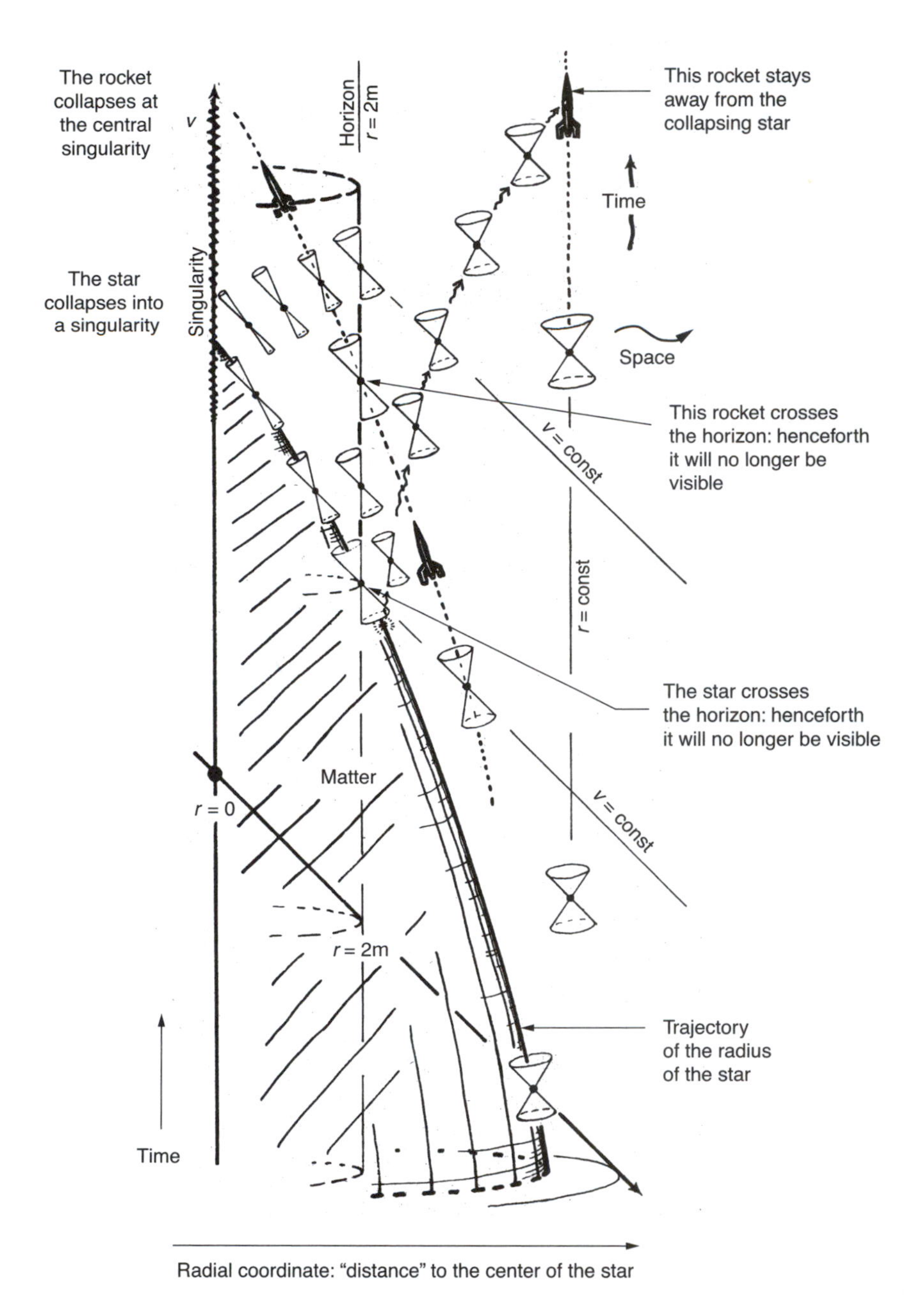

Figure 13.8. Another representation of the collapse of a star. (Based on Penrose, 1969)

the moment when a large section of the relativist community rejected this interpretation of Schwarzschild's solution. It is not enough merely to dismiss outright those conservative physicists or to cheer along with the "revolutionaries." General relativity was going through a true revolution!

We are so used, and have been for so long, to inhabiting a Euclidean space having the form of our little rooms and our plains that it is extremely difficult to imagine, to accept, that the universe may not be infinite but, rather, shaped like a saddle, and that the gravitational field of a star may hide the strange structure of a convoluted space-time.

Concerning the other regions of our diabolical diagram, in particular the white hole, the speculative character of those topological conjectures cannot be denied; conjectures which, for the moment, are neither justified by any physical reality nor predict any phenomenon susceptible of being verified by observation. But, even if they are not significant from an (astro)physical point of view, speculations such as these contribute to testing the consistency of general relativity. They represent the first stage in the construction of a theoretical model that shows the potential of general relativity. It is a sort of freedom of thought granted to scientists, those children-at-heart, so that they may have fun building highly hypothetical models on which some heavy mathematical machinery will be developed. A game, certainly, but one that is essential for a better understanding of the theory. Since then, many other "games" were put on the market, topological games whose primary interest is to liberate different forms of theoretical creativity in general relativity. Roger Penrose's works, developed by Brandon Carter of the Cambridge school,[2] are probably the best example of the possibilities offered by that freedom of thought.

The so-called Penrose diagram (figure 13.9) proposes a kind of topological schematization of the main regions of Schwarzschild's space-time. Our space-time is here reduced to its essentials! As is clear from the diagram, distances are not respected at all, for infinity is represented at a finite distance but, on the other hand, we can see better what is going on. The position of

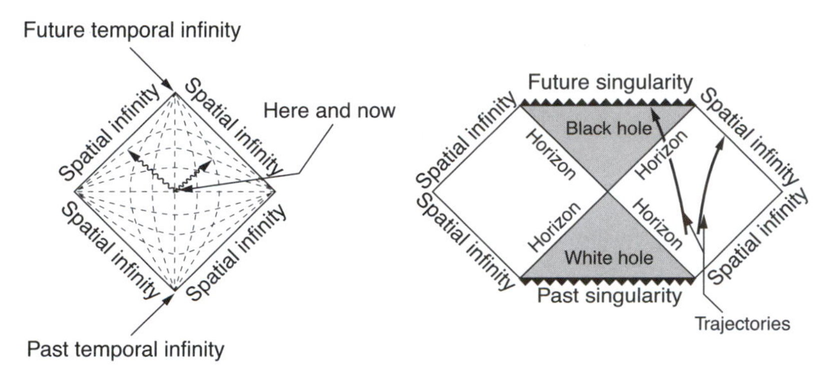

Figure 13.9. Penrose diagrams. Left: Minkowski's space-time. Right: Schwarzschild's space-time. (Based on Luminet, 1987)

the different regions of space-time is preserved, and so is the diagonal direction of the trajectories of light, but that is all. The same trajectories as in figure 13.5 are represented, together with Penrose's diagram of Minkowski's space-time. These "Penrose maps" or new diagrams of space-time may help to interpret Kruskal's diagram. That is precisely one of the roles of ongoing investigations, in addition to facilitating the emergence of theoretical speculations. An example of this is figure 13.10, a schematic representation of a Kerr black hole due to Brandon Carter. Space is divided into zones that are very strange parts of the universe. But what is a Kerr space-time?

Schwarzschild's solution determines the structure of the simplest stars, those that are perfectly spherical. But what if the star is not completely spherical (as is, in fact, always the case) or if it rotates and is embedded in a material environment, the cosmos (this also being always the case)? From a topological point of view, the problem then becomes infinitely more difficult and, given its analytical complexity, the exact solution is generally unknown. Kerr's solution, an approximate representation of a rotating star, is the simplest after that of Schwarzschild's. It is also another example of the complexity of general relativity's interpretation. In figure 13.10, there are more zones, their interpretation is trickier, and the resulting diagram is of course more complex than Kruskal's.

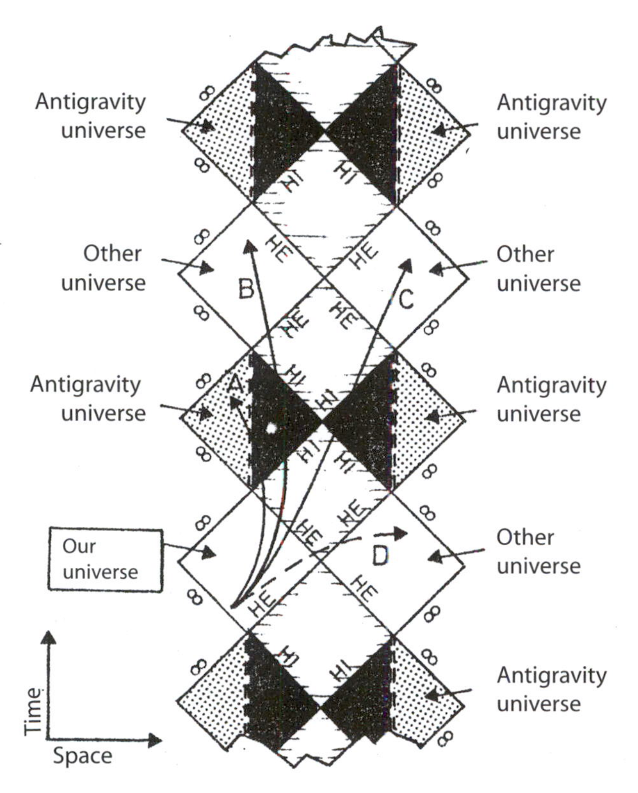

Figure 13.10. Penrose diagram of Kerr's space-time. (Based on Luminet, 1987)

Kerr's theory predicts the existence of black holes much more complex than those of Schwarzschild's: black holes that may be electrically charged and even rotate. But that is all: for essentially mathematical reasons (they are after all mathematical objects!), black holes cannot have any properties other than mass, charge, and angular moment, and these can only manifest themselves through their gravitational field. According to general relativity, the most general model of black holes can only have three parameters, so that a collapsing star, one that "crosses the horizon," will have to lose all other characteristics in order to become a black hole. Therefore, all black holes are alike and they are extremely simple objects (if we may say so . . .), at least from a physical standpoint, especially if we compare them to the familiar stars, which are so diverse from an astrophysical point of view. Such a "theorem" (the above proposition has been proved in the context of the theory) has a very powerful consequence: stars imploding toward the black hole state must, as they go through the horizon, lose all their attributes (apart from our three parameters: mass, charge, and angular moment), such as their prominences, their asymmetries, and their magnetic field; they must, one way or another, become spherical or "bald." Prior to the star's crossing of the horizon, this lost structure must be evacuated as radiation, in the form of gravitational waves that astronomers expect to detect. That is a question we shall take up again in the next chapter.

The question of the usefulness of such diagrams, of such investigations, naturally arises. Insofar as they are clearly speculative and observation plays no role, what's the point of continuing in that direction? The answer is: because those speculations are useful to ensure that the theory is viable, that it is built on solid ground, and also to find out under what conditions our interpretation is valid. In short, we should not demand too much from those diagrams except the reassurance that our current interpretation—as a black hole—is consistent. All of these reasons are related to the fact that general relativity involves a global interpretation of space-time and that we cannot seriously apprehend one of its parts without having a pretty good idea of the whole. Without those foundations, the theory would collapse like a house of cards.

In the meantime, this explanation should suffice: the purpose of the diagram in figure 13.10 with its strange extensions is to guarantee the consistency of that small piece of map on the right of figure 13.7 that is the subject of serious work; a region of space-time whose limits we want to probe in order to better understand those of general relativity.

There is no good in going on and on about it. . . . We have to accept the theory as it is, even if it is incomplete (which is partly true) and find a representation for it. That does not necessarily mean that the space we choose as the representation is true or real or even realistic; it does not mean that it exists, here and now, as certain science-fiction films would like us to believe, for it certainly does not. That does not mean, though, that the first part of the game, the black hole hypothesis, is absurd, as certain inflexible minds have preferred to believe. It means that the most pressing problems have been solved, but not everything in the garden is rosy. But what theory can be taken to its limits, to its completion, without some deceptions along the way? Only a *true* theory can. And general relativity is only *correct* within a domain of validity whose boundaries are not yet well defined. We have to be patient and accept the weirdness of this bizarre structure until something better is found—another interpretation or another theory which will not necessarily pose the same problems.

It is not even certain that all the above considerations will resist the deeper analysis that some theoreticians are presently carrying out, or that a complete interpretation of those diagrams will ever be found. Moreover, no theory is immortal, and for quite some time there has been speculation about the next theory during the wait for the demise of general relativity. For beyond a science that is "normal," beyond a theory that works well and resists all its challenges and the passing of time, a number of relativists are waiting for the moment when at last they will be able to work differently, when they will not have to be satisfied with proposing some new effect or appending one more decimal place to some known one. They are waiting for the fall, the moment when the theory will no longer agree with experience, when they will be able to offer their services and—why not?—take the father's place. One can always dream. Alas! It still endures, they think, after almost one century. They lie in wait for their time to come. That's also part of the game.

Going back to our analogy with the representation of the Earth, let us imagine a large sheet of paper (representing the universe) that is crumpled, folded, and even torn or with holes at some places, such as relief maps, which could be glued to a globe. Here, we may find an almost spherical part (a kind of nipple in the shape of a hill), there a conical element (a peak), somewhere else a plane section (a plain, of course), there a hole (a precipice or a lake), with the whole thing, the universe itself, lying on some spherical surface, just like our globe. This is the task that needs to be accomplished and that we are only beginning: how to represent a star and how to connect the geometry of its gravitational field to that of the universe.

Recall that at each point of the universe, space-time appears locally flat, with no curvature, with the speed of light equal to c, and special relativity applies. This is what is usually called the "tangent space" or even the "tangent plane," a name that indicates the extent to which the geographical metaphor is appropriate (see chapter 12).

DID YOU SAY "BLACK HOLE"?

The history of the name given to the notion of black hole reflects very well the mistakes, hesitations, and fears of theoretical physicists throughout the century, and even their desire to sell their product. From the 1920s to the 1960s, the relativistic literature is full of those more or less technical or more or less colorful expressions, their number growing as the object becomes more disturbing. Each choice has its reasons, its history, and its context.

The choice of an expression is anything but innocent and it betrays its author's intentions: to lead the reader toward a certain interpretation, a certain vision; to favor a certain way of looking at the thing over some other. The point is important, for the name given to an object or an individual will necessarily have repercussions on their future life. It may be, first of all, a slip, insofar as the choice of a name or nickname unintentionally betrays the way one sees the object, the person, or the child. But it is also a means of trying to influence the future. The connotation of a word or a name is far from being harmless, even if it is sometimes unconscious. A somewhat ridiculous nickname or a first name with social implications will not go unnoticed. In short, there is nothing more important than making the right choice of a child's first name, a pen name, or a stage name—and, for the physicists, of the name of a concept, instrument, or idea.

They are well aware of this, and a paper could be written on the choices that were and are being made, on their reasons and the strategies they conceal or imply, and on the influence they had and still have on the evolution of the object, the concept, or the tool.

At the end of the eighteenth century, John Michell, who had just invented a kind of photometer to compare the magnitude of stars, was looking for a name for his instrument. He observed that "a hard name adds much to the dignity of a thing,"[3] and he wished "to take the liberty of christening an instrument for this purpose as Astrophotometer."[4] However, that well-chosen name did not benefit either the instrument or Michell (a few drops of holy water are not enough to go to heaven), and Michell's splendid work on that subject (and on others) was completely ignored and forgotten.

Michell's idea of those strange stars whose light is trapped by their own gravitational field was recovered by Laplace a few years later in his *Exposition du système du monde*, but not without christening them (specifically, "dark bodies"), something Michell had omitted to do. But naming them did not prevent the dark bodies from being forgotten, too, probably because at that time they had no reason to exist; however, Laplace had nevertheless made the concept invented by his colleague Michell his own.[5] It is clear, as his treatise shows, that Laplace had a talent for writing that Michell singularly lacked. The Englishman was no doubt a great physicist but a bit of an introvert. That is certainly not the case for John Wheeler, who invented the expression "black hole."

The question of naming that strange place, the horizon, and the object it hides had come up at the time of Schwarzschild, who employed the term "discontinuity." If you speak of "singularity," you will be necessarily stuck, blocked, and the idea of impenetrability will immediately enter your mind. You will not be able to think of that place other than as a forbidden one. And this is exactly what happened in the first half of the century, when the expression "Schwarzschild's singularity" was everywhere: in textbooks, in classrooms, and in minds.

Recall that Einstein saw in that place—on the horizon—a catastrophe, "Hadamard's catastrophe," an idea present also in France and Belgium in the expression *sphère catastrophique*. This catastrophic angle was taken to extremes by certain authors who introduced the idea of death (the expression "dead point" was used, or even a sanctimonious "death"), probably because time was believed to come to a stop there. As we have seen above, there was indeed a catastrophic effect at the real singularity (discovered much later), and for better reasons. The expression "magic circle," due to Eddington, was used at the time by many authors, as well as "limit circle" and "singular sphere." A number of terms, such as "frontier" and "barrier" would refer in particular to the impenetrability then believed to be absolute.

After the 1960s, the topological interpretation produced an evolution of terminology paralleling that of the interpretation itself: the expression "wormhole" appeared, and also *terrier de garenne* (rabbit hole), and even "throat of the Schwarzschild's solution." The accent was on the matter, energy, and light hidden behind or on this frontier, in such terms as "sink of photons" and "matter horn": the idea of a distorted space was beginning to be accepted, opening the way to a topological interpretation.

The expression "black hole" did not appear by chance, and its proposal by John Wheeler was anything but uncalculated. He was pictured complacently lying in bed, looking for the right word, because he did not like the terms "frozen star" or "collapsed star" then in use. After he found the right expression, at the end of 1967, and in order to impose it in the most effective way, he employed it "as if no other name had previously existed." He not only described what a black hole was but dressed it up in a profusion of images: "What was once the core of a star is no longer visible. The core, like the Cheshire cat, fades from view. One leaves behind only its grin, the other, only its gravitational attraction. Gravitational attraction, yes; light, no."[6]

The name thrived, and nobody could deny the impression made—inside as well as outside the discipline and also in the popular science literature—by this strange expression, the "packaging" of a new and subtle concept; a choice also motivated by its marketing effect and by its contribution to the "great science show."

Considering that all images are inexact, many scientists are reluctant to use them. Wheeler, on the contrary, was not afraid to employ a variety of them, for he probably—and rightly—believed that a false image is better than no image at all. But don't we have equations to reestablish the true meaning of the concept? Yes, but it is important that the mind should get a foothold on the idea if it is to move forward, for equations are a priori too obscure or too transparent for the mind to consider them. It is important that words and the images they suggest should imperceptibly guide the mind toward the essence of the concept. But before christening a concept one must first know what it means.

NOTES

1. On this point, cf. Eisenstaedt, 1982, p. 167–171.
2. In this regard, cf. Hawking and Ellis, 1973.
3. Quoted by McCormmach, 1968, p. 129.
4. Michell to Cavendish, 2 July 1783, in ibid.
5. In this respect, cf. Eisenstaedt, 1991, 1997.
6. Wheeler, 1968, p. 9.

CHAPTER FOURTEEN

No Ordinary Stars

IN THE PRECEDING CHAPTERS we have examined the structure of space-time in the vicinity of a spherical mass. As we have seen, that structure implies the existence of black holes, but can such objects physically come into being? This is the first question we will address: the evolution of stars whose final stage of development is, or would be, a black hole—provided we can see them, because they are black. . . . In a word, have black holes ever been (or may soon be) observed?

Before discussing the dynamics of the explosion of a star (when that happens they are called novae or supernovae) or of their collapse (they then become white dwarfs, neutron stars, or black holes), let us see how an ordinary star can be, at least for a while, in a state of equilibrium.

Let us begin with the familiar example of our own Earth, on which we are, without being aware, in equilibrium. This is a consequence of Newton's action-reaction principle: the forces involved must balance each other out because we *are* in equilibrium. Basically, the force of gravitation (our weight, which is a function of the mass of the Earth) is balanced by the pressure exerted by the surface layers of the Earth. And not only its surface, but each layer, each piece of each layer, at every depth exists in equilibrium between the gravitational force acting on it and the opposing local intermolecular forces. If the latter are not strong enough, the soil collapses, as it does occasionally when the surface layers no longer balance the load, for reasons related to the physical composition of the ground; a local collapse then follows which will stop at a more resistant layer. If at a certain spot the pressure becomes stronger, due to an increase in temperature, for instance, an explosion follows, as in a volcanic eruption. We must realize that this equilibrium is fragile: we actually live on top of a volcano that is calm most of the time.

In stars, these phenomena are more violent, more complex, and extremely varied depending on the type of star. I will focus on the fundamental mechanisms, in particular those related to gravitation, that will allow us to have a rough idea of what is going on—or what astrophysicists think is going on, because they often have to resort to conjectures that require theoretical and empirical confirmation.

There is an essential difference between electromagnetic and gravitational phenomena: while there are both repulsive (the negatively charged electrons) and attractive (the positively charged positrons and protons) particles, in gravitation there are no repulsive masses. The force of gravitation is cumulative, whereas the electromagnetic forces cancel out at short range and the local electric field is soon neutralized. This is why the gravitational field manifests itself at long range. On the surface of the star, the gravitational field may also increase when the radius of the star decreases while its mass remains constant. Therefore, the increase in gravitational force may be due to either an accretion of matter or a contraction of the star. Needless to say, gravitation plays a crucial role in star formation. It is believed that at the very early stages of the process, interstellar clouds contract as a result of a gravitational accumulation of matter; this contraction increases the density of the gas. The frequency of the collisions between particles also increases and, as a consequence, the temperature of the star rises; a proto-star forms and later contracts to become a star. A star's life span is principally a function of its mass, those having smaller masses living longer.

THE STAGES OF A STAR'S COLLAPSE

At the end of each dynamic process—contraction or explosion—each layer of the star is in an essentially stable equilibrium state. As in the case of the Earth, two types of forces are at play: on the one hand, the gravitational forces striving to bring about the collapse of the star; on the other, the reaction or pressure forces (which depend on the star's composition, density, and temperature) that counterbalance the gravitational force. It is this struggle between gravitation and pressures that explains the star's dynamics. If the gravitational forces prevail, a new stage of the collapse begins. But if pressures win out, the star explodes. Equilibrium is only a temporary state in a perpetually dynamic process.

A star progressively accretes, sucks in matter: its mass increases and so does, at first, its radius. Its gravitational force, proportional to M/r^2 (in Newtonian terms) increases too, and if the process does not stop, the gravitational force at the surface will become stronger than the cohesion forces; the latter remain constant, for they depend only on the composition of the star. The increasing gravitational forces produce an initial collapse which, in compacting the material making up the star, results in a new equilibrium. During this process the radius of the star decreases, so that the gravitational force (proportional to M/r^2) increases.

But another phenomenon also contributes to the collapse: the star radiates, and that requires energy. During a significant part of its life, the center of the star emits heat from the thermonuclear fusion that transforms hydrogen into helium. Its temperature reaches 10 to 20 million degrees, and this thermal pressure prevents the implosion of the star under the effect of its own gravity. Eventually, the star will run out of fuel and its temperature, and hence the pressure, will decrease; but not gravitation. A new collapse then becomes inevitable. Such is the star's infernal cycle: it cannot escape collapsing; it is its destiny. Let us see how far their destiny takes them.

A hot star such as our Sun will necessarily evolve. The Hertzsprung-Russell diagram classifies stars relative to two basic parameters, mass and luminosity, so that we can *follow* their evolution using this diagram. Roughly speaking, there are three possible types of evolution for an ordinary star depending on its initial mass, which determines what type of star will result after a period of instability:

a white dwarf, if the initial mass is less that one-and-a-half times that of the Sun (this is what our Sun will become);
a neutron star, if the initial mass is between 1.5 and 3 solar masses;
a black hole if the initial mass is larger than three solar masses (approximately).

The last is a clear possibility, for there are stars whose mass would be one hundred million times that of the Sun but with the density of water. It must be added that if, at the beginning of its evolution, the star has an extremely large mass, it will go through a period of instability of the supernova type—that is, an explosion during which it will lose most of its mass by ejection of matter—before it falls into one of the above categories. But what kind of "stellar material" is present in each of these three cases? With the help of quantum mechanics and nuclear physics, astrophysicists have answered this question.

In the case of a white dwarf, the gravitational force is counterbalanced by the pressure of the degenerate electron gas, basically, the impenetrability of electrons combined with Pauli's exclusion principle (the fact that two fermions cannot be in the same quantum state). These stars, whose typical mass is approximately equal to the Sun's, have a radius comparable to that of the Earth's and a density of the order of one ton per cm^3, that is, a million times that of the Earth or the Sun (which is about one gram per cm^3). We have seen in chapter 9 how, through the Einstein effect, general relativity helped to understand white dwarfs.

As for a neutron star, the gravitational force is counterbalanced by nuclear forces, the pressure of the neutron gas, which is also related to Pauli's exclusion principle. The typical size of these stars is ten kilometers and their density is approximately 10^{14} g/cm^3, that is, more than one hundred thousand billion times denser than the Earth. However, no "pure" neutron stars have ever been observed; only another, more complex kind of object: neutron stars in extremely rapid rotation known as "pulsars."

And then what happens to the star? No known force can prevent the implosion, so the star will completely collapse until its radius becomes zero. The gravitational force on the surface of the star can increase indefinitely in two different ways: if the mass M creating the gravitational field increases or if the radius R of the star decreases; for the gravitational force is proportional to M/R^2, and may even tend to infinity if the collapse continues all the way to the center. However, the cohesive forces of the fluid making up the star—the quantum and nuclear forces—are finite and depend on the type of stellar material and the state of matter. This is why gravitation will always prevail over cohesion. And then, nothing, absolutely nothing, no reaction force, cohesion, or pressure from the star's gas will be a match for the gravitational force. The star will be in "free fall," a collapse nothing can now stop. But first, the star will have to traverse its own horizon and become a black hole.

AN INESCAPABLE FATE

For a black hole to form, the collapsing star has to cross its own horizon. The ratio between the horizon's radius and the star's radius is $2GM/Rc^2$; this is the characteristic value (see box 5, chapter 7) that indicates whether or not the star is approaching the black hole state. If this value is very small, we have a Newtonian situation. The value increases with the effects of general relativity, and when it is close to 1, the star is near its horizon, which it crosses if $2GM/Rc^2 = 1$. At that point, Schwarzschild's horizon is no longer potential but becomes real and prevents all particles from leaving what is left of the star. The trap is shut. The escape speed is now equal to the speed of light. Even the light particles can no longer free themselves from the gravitational field; all particles of the star are trapped inside the horizon. The black hole is formed, defined by its horizon.

A Schwarzschild's horizon is a strange surface. While the surface of an ordinary star is full of turbulence—jets, various kinds of spots—and not quite spherical, the horizon of a black hole is a perfect sphere, a sort of "mathematical" surface or bubble with no imperfections, made up of light particles "trapped" by gravitation. Here, mathematics does more than merely model physics; it imposes its shapes upon it: the shapes of the pathways of light.

To put things into perspective, consider the following: the radius of the Earth's horizon is approximately one centimeter, while its actual radius is a little more than 6,000 km. This means that for the Earth to be a black hole, all its matter would have to be contained in a sphere of radius one centimeter, that is, the whole Earth would have to collapse to the size of a thimble. As for the Sun, the radius of its horizon is about three kilometers (while its actual radius is close to 700,000 km), the same as the radius of a white dwarf's or a neutron star's horizon—their masses (but not their radii!) are comparable to that of a conventional star such as the Sun. The radius of the horizon ($2GM/c^2$) of a star is proportional to the mass of the black hole it determines.

But, is the formation of a black hole the only alternative? Does general relativity really predict the existence of black holes? One might think—as was the case in 1920–1930—that such a thing just "could not happen." A surprising argument was developed according to which for a black hole to form the density of the contracting star had to be so enormous that it was thought physically impossible to attain. However, as we are going to see, that is not necessary. In order to convince ourselves of the inevitability of the existence of black holes, let us have a look at the conditions under which a horizon, or black hole, may form.

We are going to use a Newtonian force, which is easier to manipulate even if it is not entirely correct from a "relativistic" point of view. At a distance r from the center of a star, the gravitational force is proportional to M/r^2 (we are neglecting here all constants and keeping only the variable quantities, M and r); the radius of the horizon is proportional to M (it is equal to $2GM/c^2$), so that the force *at* the horizon is proportional to $1/M$, that is, inversely proportional to the mass. Hence, the greater the mass of the black hole, the smaller the gravitational force at the horizon. Similarly, the density[1] is proportional to M/r^3, and, given that the radius of the horizon is proportional to M, the density of a star that "crosses" its horizon is proportional to $1/M^2$; hence the greater the mass of the star, the smaller its density (inversely proportional to the square of M), at least at the moment it crosses its horizon. This means that a star may "burst" its horizon, that a black hole may form, if the mass of the star in question is large enough; and the larger it is, the easier for us to envisage such a possibility, for then there is no need for a high density. Thus, the density of a black hole in the process of being created may be extremely low, an ordinary density. We no longer need ultradense matter, even though shortly after the crossing of the horizon, the implosion becomes inevitable and, in almost no time, shortly after its disappearance—that is, after traversing its Schwarzschild's horizon—its density becomes infinite. And so black holes of very large mass may be created with very low density, but the latter has to be offset by their size. In short, one way or another $2GM/Rc^2$ has to be greater than 1.

A star whose mass exceeds three solar masses is therefore doomed: it cannot reach equilibrium. Inevitably, a black hole will form the moment the star crosses its own horizon. And each one of the collapsing star's particles goes through the horizon unharmed. At this moment, all communication with the outside world ceases. The collapse continues and all the particles of the star, pulled by those which preceded them, converge toward the center. While the imploding star tends toward the point (a truly singular one) $r = 0$, the origin of space-time, the density of the particles approaching the center grows without bound. The gravitational forces increase until they become infinity, and the density of matter is such that quantum forces necessarily play a role, one that general relativity cannot account for. But at the very center of the star, the density becomes infinity, which, properly speaking, has no physical meaning, and the theory itself becomes singular: it simply no longer applies.

This phenomenon of implosion toward the center of the black hole is precisely the same one we shall meet (in the next chapter) when describing the big bang. It is the same thing, but in the opposite direction; space-time will explode from a singular point (in space-time), and from that moment on it will expand. As one might easily imagine, this peculiar origin of the universe, like the singular fate of stars, creates real problems for theoreticians.

More precisely, a limit of general relativity has been reached, for the theory is meaningless in those strange places that do not even belong to the space-time whose equations define the theory. A "singular" point (where the components of the fundamental tensors, the intrinsic magnitudes, are infinite) cannot be a point in space-time. It is impossible to tell what happens to a trajectory that reaches such a point, one where space-time is not defined. A point is missing, and space-time is therefore not "complete." This is a serious problem for general relativity: it has encountered a mathematical limit, a limit of the very structure of the theory. A unified theory is therefore called for, one that will integrate both the quantum and the gravitational fields. Gravitation itself may have to be quantized. I will elaborate on that in chapter 15.

A black hole has been formed. What kind of phenomenon awaits us at the horizon, at its boundary? In order to understand the questions related to energy, the driving force of all of physics, let us bring another star close to our black hole, a kind of test star, much like we would bring a midge near a spider's net. We want to study the system composed of a black hole and an ordinary star that will soon be captured, distorted, torn up, and finally swallowed by the black hole. Needless to say, it is an extremely complex problem, and we shall only attempt to take a look at what may happen.

Let us follow the star in its infernal fall. . . . First of all, nothing special will occur as the star crosses the horizon, except that it will disappear from the sight of an external observer. But, as the star (i.e., what remains of it) approaches the center of the black hole, the curvature of space-time (or, simply put, gravitation's pull on the star) will continue to increase until it becomes infinite. The gravitational force or curvature of space-time is stronger for parts of the star near the center of the black hole than for other, more distant portions. This gives rise to a gravitational force difference acting on the star as it falls into the black hole. These are similar to the tidal forces that in Newtonian theory account for terrestrial tides, a result of the gravitational force difference that the Sun and the Moon exert on the oceans (as well as elsewhere on Earth). Just as the Earth is distorted by the joint action of the gravitational fields of the Sun and the Moon, our star, subject to the gravitational field of the black hole, will gradually deform and take the shape of . . . a cigar. But, as it keeps falling toward the center, the curvature of space-time grows and the tidal effect (which is related to the local variation or gradient of the gravitational filed) becomes relentless. Subject to such tremendous constraints, our star will be torn up, and its pieces will undergo extremely violent phenomena. These events only take place beyond the so-called Roche radius of the black hole.[2] However, they may occur either inside or outside the black hole, depending on the Roche radius: if its value is smaller than the horizon, nothing will be seen from the outside; but if it is greater, these exotic phenomena will take place as soon as our star approaches the horizon, and they could allow us to visualize, to identify the black hole.

"GORE AT THE SINGULARITY"[3]

In order to illustrate these phenomena, the authors of *Gravitation* have described one of their colleagues, an astrophysicist, falling into a black hole.[4] If we wish to really understand what happens to a star falling into a black hole, let us follow their colorful account.

This piece, which reads like one from an anthology of crime literature, will also reveal the effort made by the authors to provide their students with images of the situation. Their goal is to shake the theory out of its torpid condition, which has now lasted for too long. They want to strike hard and catch the student's imagination, and also that of the relativists, who too often still regard Einstein's theory as a "quiet backwater of research."

Gravitation is an immense book, almost thirteen hundred pages long, that students simply call "the phone book." Published in 1973, it illustrates with elegance the renewal of Einstein's theory. Black holes are the central theme of the book, a topic the authors know firsthand from their own work. All possible representations of a black hole are presented, including some invented for the occasion, historical (in particular, references to Laplace's dark bodies), technical, and literary ones. No effort is spared to illustrate a theory badly suffering from a lack of images and of visual representations.

To best illustrate the tidal forces in action near the central singularity, they describe in the chapter on gravitational collapse "the fate of a man who falls into the singularity at $R = 0$." Three pages are devoted to this event, which more than any other catches the imagination and illustrates the theory; three pages,[5] one of which begins with the gruesome heading "Gore at the singularity." And, of course, it is an experimental astrophysicist (the authors are theoreticians; isn't there some unconscious hostility here?) who bears the consequences of this thought experiment:

> Consider the plight of an experimental astrophysicist who stands on the surface of a freely falling star as it collapses to $R = 0$. As the collapse proceeds toward $R = 0$, the various parts of the astrophysicist's body experience different gravitational forces. . . . In order to accomplish this, tidal gravitational forces must compress the astrophysicist on all sides as they stretch him from head to foot . . . ; so the astrophysicist, in the limit $R \to 0$, is crushed to zero volume and indefinitely extended length. The above discussion can be put on a mathematical footing as follows. (860)

The mathematical model occupies two pages covered with formulas, and it is divided into the three stages in the killing of the astrophysicist: "(1) the early stage, when his body successfully resists the tidal forces; (2) the intermediate stage, when it is gradually succumbing; and (3) the final stage, when it has been completely overwhelmed."

These gruesome details are not (only) meant to amuse some sadistic colleagues or readers, as a superficial analysis might suggest. Each stage illustrates a different physical effect: the fall toward the black hole, the crossing of the horizon, and the approach to the true singularity. The realism of the account extends to the choice of the physical units: pressure forces are expressed in dynes/cm^2 or in millibars, that is, on a human scale. It is a realism that stems from the authors' concern for experience, for the applications of the "true" physics. It was a perfectly justified concern: wasn't general relativity up till then considered too abstract a theory, with too few applications? The authors' purpose is to develop in their young colleagues a feeling for orders of magnitude.

Let me make one thing clear. This is—can only be—a thought experiment. In this type of experiment, and for reasons that have nothing to do with being humane, there are limits: it is an experiment that can never be performed. Even if we had a black hole and a consenting colleague, putting them together would not serve any useful purpose, because no communication from our astrophysicist, no account of what happened, could ever reach us. We would have to go ourselves, and we could only verify our hypotheses up to the second stage. What we have here is, therefore, a conceptual show, a thought experiment not meant to be a test but only an example, a theatrical representation of the theory whose purpose is to render the theory, its concepts, and models more concrete; in a word, to "put some flesh" on the ideas of the theory.

A few pages earlier, these professors had proposed to their students an exercise on the same theme: find the mass of a star which "a normal flesh-and-blood human being" could endure. Get that into your heads. At all costs!

THE HALLMARK OF A BLACK HOLE

In a strict sense, a black hole is optically unobservable; only its gravitational field is, beyond its horizon, perceptible. This does not mean that there are no induced effects, far from it. But how can we predict the characteristic effects of the presence of a black hole, that is, its "hallmark"? How does a black hole manifest itself?

Obviously, we shall never have any news of our unfortunate colleague or of his disintegration; his sacrifice would have taught us nothing about the structure or even the existence of that black hole. The same is true of a massive black hole which, given that its gravitational field at the horizon is weak, may swallow stars before they can be dismembered by the tidal forces. In that case, the star will disappear into the black hole without emitting any high-energy radiation; the catastrophe will go unnoticed. On the other hand, if the mass of the black hole is smaller than 10^7 solar masses, the star will break down before being swallowed. Something remarkable then happens near the black hole, something so violent in terms of energy that it is believed to be one of the most dynamic stellar mechanisms—one that might allow us to understand what we observe and predict what will follow.

This explains today's interest in double sources or *binary* systems, which consist of a compact object—a neutron star, pulsar, or stellar black hole—around which an ordinary star revolves. The compact source then generates a powerful gravitational field and strongly relativistic effects that the companion star allows us to test. Such is the case of PSR 1913 + 16, a quite extraordinary double system, formed by a neutron star around which a pulsar is orbiting. It is a highly dynamic and extremely precise system, an ideal laboratory to put general relativity to the test, for it exhibits a remarkable advance of its perihelion.

If the star happens to be a bit too close to a black hole of small mass, it will be torn to pieces by the tidal forces before disappearing into the black hole. This drama in the life of stars, to which astrophysicists specializing in relativity have devoted sophisticated models, seems to unfold in several stages and, in particular, from an intermediate state consisting of a disk orbiting around the compact object—an "accretion" disk, similar to Saturn's rings. The black hole's gravitational field gradually sucks in particles, and the gas in the disk becomes increasingly hot. Soon the temperature exceeds ten million degrees, and energy is released in the form of electromagnetic radiation, X or γ, that astrophysicists hope to be able to observe. Closer to the horizon, the structure of the phenomenon becomes extremely turbulent, and the star's debris is absorbed by the compact object, a complex phenomenon on which I will not elaborate.

Another scenario involves a revolving black hole losing rotation energy. If a particle enters the gravitational field near the horizon, it breaks into two parts, one of which is absorbed by the black hole, while the other is ejected, carrying away a fraction of the black hole's rotation energy. The fusion or coalescence of two black holes is another much studied possibility: before they merge, there should be a significant emission of gravitational waves.

In short, for a black hole to reveal itself, it must be combined with some other object: particles, ordinary star, neutron star, or another black hole. Its silence is a consequence of the perfect sphericity of its horizon, and it is only by upsetting this symmetry that we can expect to make it speak to us.

Naturally, all these scenarios are only models (presented here in an extremely simplified form) from which significant elements may be gleaned so that a distant observer might be able to detect the presence of black holes, a detection that can only take place indirectly, by the observation of secondary effects on the gravitational field of the compact source. A particular radio-, X, or γ radiation may indicate the existence of a black hole and constitute, in a certain sense, its hallmark.

Therefore, "to observe" a black hole means to detect radiation induced by its gravitational field. But not all induced radiation will signal the presence of a black hole; one must make sure that the radius of the source does not exceed a few kilometers, otherwise it may be a neutron star. If the radius of the source is about three times that of the horizon, it is "only" a neutron star; and if it is much larger (typically three thousand times larger), we would be in the presence of a white dwarf. But if our object is both invisible and with a mass larger than that of a neutron star, we may then conclude that we have indeed spotted a black hole.

The truth is that, to this day, specialists have reached no consensus on this question. Most of them are "absolutely convinced" that black holes exist, that it is highly likely that some of the "candidates" are really black holes, and that soon everybody will be convinced, too. But, for the moment, the doubt persists and certainty is far from being complete, despite what certain astrophysicists loudly proclaim. Isn't it imperative to be the first one to have observed a black hole, to be the one who will discover the object that will convince the entire community?

The list of objects believed to be likely candidates for a black hole is not very long, and many of them have actually proved to be neutron stars. However, once again, it is not impossible that there are true black holes among the objects on the list. All we can honestly say is that there is no general agreement, but at best a serious presumption, that this or that astrophysical source is a black hole. Insofar as one is convinced that general relativity predicts the existence of black holes—and on this point specialists are unanimous—one may reasonably expect to observe the indirect effects of such strange objects anytime now.

Obviously, the fuzziness surrounding the possible observation of black holes opens the door to a variety of points of view, degrees of certainty, and dubious strategies. Hardly does a month go by without some reference to black holes somewhere (and even more so in the popular press), but these expectations have so far been thwarted.

The observational criteria for a black hole generally include the presence of a strong X-ray emission, the visual identification of a companion, and a sufficiently accurate determination of the compact object's mass through the study of its companion star's velocity. In fact, only a few binary black holes are presently being studied, in particular Cygnus X1, which is still only a "good candidate" and was already mentioned in *Gravitation*'s foreword, published in 1973. It was in 1973 that the "black hole" project exploded in the literature with the publication of several books on general relativity featuring the notion of black hole, which then became a driving concept in general relativity. In 1978, it was M 87's turn, an elliptical galaxy of gigantic proportions situated in the Virgos cluster. Since then, new candidates have been added: LMC X-3, an X-ray source whose mass would be larger than five solar masses, and the nova 0620-00, a bright source of X-rays. Not long ago, a black hole was detected at the center of our galaxy.[6]

NOTES

1. Simply because in Euclidean geometry the volume of a sphere is equal to $4/3\pi r^3$. Since the volume of a sphere depends on the geometry

of the space, this calculation would not be correct in general relativity because in it the geometry is not necessarily Euclidean.

2. The Roche radius is the distance such that at any smaller distance a moon is destroyed by the tidal forces.

3. This is the macabre title of section 32-6 of the 1973 book by C. W. Misner, et al.

4. Misner et al., 1973.

5. Ibid., pp. 860–862.

6. In this regard, see Israel, 2000.

CHAPTER FIFTEEN

Gravitation, Astrophysics, and Cosmology

THE EXOTIC OBJECTS OF THE 1960s

THE 1960S MARKED the development of high-energy astrophysics. Numerous dynamic phenomena and new astrophysical objects were discovered that would constitute general relativity's preferred hunting ground.

The reason general relativity had practically no applications before the 1960s was the almost total lack of an object huge, massive, or dynamic enough for the "relativistic terms" describing its gravitational field to be of any interest, since such terms remained "negligible." It was in fact a theory well ahead of its time and of its observational domain. In a ten-year period, astronomers discovered a large number of phenomena and stellar objects which, one way or another, would enormously help the understanding (and verification) of Einstein's theory: quasars, pulsars and other neutron stars, black holes, and the cosmic background radiation at 3°K.

Quasars are extremely distant and extremely bright objects. They could be nuclei of galaxies, nuclei brighter than our own galaxy, made up of an ultramassive black hole swallowing hundreds of millions of solar masses; an "engine" whose dynamics would be based on the conversion of gravitational energy into radiation with a 40 percent performance rate. But for now this only a (serious) working hypothesis that not all specialists accept.

In 1967, at Cambridge University, a young radio astronomer was working on radio signals emitted by quasars. She detected a very strange radio signal: periodic electromagnetic impulses of very high intensity. They had nothing to do with the signals from the only variable stars known at the time, the Cepheids, which were infinitely less dynamic and whose period was much longer. These new signals were more regular than our best atomic clocks, and their period was so short that astronomers wondered for a while whether they might come from some extraterrestrial civilization. How could those signals be explained? It was impossible to interpret them as the radial vibration of a star or as the rotation of a star around another, for their period was much too short. Finally, the only possible model had to be accepted: the extremely fast rotation of a star whose size was necessarily very small and whose cohesion was extremely high (for otherwise it would have burst into pieces); that is, a rotating neutron star with an extremely high magnetic field, which was called a pulsar. Soon, dozens—hundreds—of pulsars would be detected, which, incidentally, shows how at a certain moment all doubts disappear and a consensus is reached, and why this is not yet the case concerning black holes.

In 1974, Russell Hulse and Joseph Taylor proceeded to systematically draw up a list of all pulsars using the Arecibo radiotelescope, in Puerto Rico. They discovered a strange source, a pulsar, PSR 1913 + 16, emitting about seventeen impulses per second. But the period of those impulses was not absolutely constant: it varied by a few millionths of a second from one day to the other, a surprising variation for a pulsar. This variation was also cyclic, with a period of seven hours and forty-five minutes, and it was caused by a Doppler effect. The pulsar was therefore revolving around a companion, which turned out to be a neutron star; an extraordinary double system, highly dynamic and extremely precise, that would prove an ideal laboratory for the study of general relativity. They also discovered that PSR 1913 + 16 has a precession motion around its companion, for the intensity of the gravitational field produces an advance of its perihelion similar to that of Mercury. But the physical conditions there are completely different from those found in our solar system, and the magnitude of this advance is 4.2 degrees of arc per year, while Mercury's is only 43 seconds of arc per century. In one day, the precession of the perihelion of PSR 1913 + 16 is equivalent to Mercury's advance in one full century. This was a magnificent result, fully consistent with the prediction of general relativity, which was confirmed in a remarkable way. These two

Figure 15.1. Crabe nebula. These are the remains of a supernova. It was observed in July 1054 by Chinese astronomers. At its center, there is a pulsar whose frequency is 30 revolutions per second. (FORS Tean, 8-2 meter VLT, ESO)

astrophysicists were awarded the 1993 Nobel Prize in physics, "for their discovery of a new type of pulsar, a discovery that has opened up new possibilities for the study of gravitation."

Einstein's theory of gravitation was finally verified, no longer only marginally in connection with its neo-Newtonian effects, but in its natural domain, that of strong gravitational fields.

GRAVITATIONAL WAVES AND MIRAGES

Thanks to their recordings over many years of the impulses of PSR 1913 + 16, Hulse and Taylor were able to show that the distance between the two stars decreased due to a loss of gravitational energy. These observations *implicitly* confirmed the existence of gravitational waves. Even though the emission power of these waves is immense—of the order of one-fifth the solar power, which is considerable—it is not strong enough to be detected from the Earth.

Gravitational waves are to general relativity what electromagnetic waves are to Maxwell's theory; the former are, in a certain

Figure 15.2. Virgo, a French-Italian gravitational wave detector under construction. Aerial view. (© Ph. Plailly/Eurélios)

way, the counterpart of the latter. Just as charged particles emit electromagnetic waves when they are accelerated, mass particles emit, when their acceleration varies, gravitational waves that propagate at the speed of light. But while electromagnetic waves travel in Minkowski space, gravitational waves are oscillations of general relativity's space-time that propagate gravitational energy.

In the 1960s, in their efforts to detect gravitational waves, astrophysicists resorted to resonant bar detectors. These are aluminum or niobium rods weighing one or two tons, which were cooled to a temperature of a few degrees Kelvin to improve their performance. The passage of a gravitational wave should alter the rod's vibration pattern. Unfortunately, the device did not work as planned. Giant interferometers (each arm was several kilometers long) using ultrastable lasers have also been built (see figure 15.2). As a gravitational wave passes through, the length of the interferometer's arms should change, resulting in a shift of the interference lines. However, the expected effect is extremely weak, of amplitude 10^{-21}. This means that, for a one-kilometer arm, the variation in length to be detected is of the order of one millionth of a billionth of a millimeter. Scientists believe (or hope) that such interferometers will be sensitive enough to detect an effect of that magnitude. Impressive precautions are taken regarding the quality of the materials, mirrors, suspension, and vacuum surrounding the beam.

The French-Italian Virgo project is under construction at Cascina, near Pisa. It is also a laser interferometer, consisting of two perpendicular arms three kilometers long. Thanks to multiple reflections, the optical path has been extended to 120 kilometers. Virgo will be sensitive to a large frequency spectrum, from 10 to 60,000 hertz. This should allow the observation of, for instance, the coalescence of binary systems in our galaxy as well as in others. The analysis of an interferometer's signals will probably not be enough to detect an event with certainty. A parallel analysis of data from other sources, such as the LIGO project presently under construction in the United States, will be needed.

For frequencies below 10 hertz, the seismic noise drowns any measurement. To circumvent this obstacle, the European Space Agency has set up the LISA spatial program, composed of six capsules in solar orbit that together form an interferometer with a five-million-kilometer span.

Since light is deflected by gravitation, it is then not surprising to find phenomena similar to those in traditional optics. For instance, the deflection of light by gravitation and the refraction of light by lenses have much the same consequences. Roughly speaking, "gravitational lenses" are formed by the accumulation of matter present in the universe. If light waves coming from a distant source pass near a very large material object, they are refracted, focused by the gravitational field, and give rise to a variety of phenomena: distorted, repeated, weakened, or even reinforced images. The deflection of light by a gravitational field is an extremely weak phenomenon (the Sun deflects a light beam less than two minutes of arc at its edge), and decades went by before this kind of phenomenon could be observed.

The first gravitational mirage was observed at the end of the 1970s. Two quasars (0957 + 561 A, B) were observed along two different lines of sight that were very close, forming an angle of less than six seconds. Quasars being very rare objects, their proximity raised a question that was quickly dissipated thanks to spectroscopy. The two quasars turned out to have identical spectroscopic characteristics (i.e., similar emission spectra), which was truly surprising. Both quasars presented not only the same set of lines, but the intensity of each line was the same for both of them; moreover all lines had the same redshift, within the measurements' margin of error. Was this a binary quasar, or only one quasar and its repeated image?

The probability of distinguishing two objects with the same spectrum in such a tiny portion of the sky was practically zero, so it was concluded that even if the images were different, they were coming from one and the same object; in other words, it was a gravitational mirage. The phenomenon was also observed using radio waves, and the galaxy responsible for the deflection was eventually found, though not without effort: it was a very pale spot between the two images, located midway between the quasar and our own galaxy. But how could this happen?

Several conditions must be met for this phenomenon to take place, but geometry (for it is ultimately a question of geometric optics) is crucial here. The deflector (typically a galaxy but it could also be a black hole) must not only be located on the trajectory of the light rays, but also at a certain specific distance from the source for the observer to receive the primary and/or secondary images. Depending on the configuration, the observer will receive a double image or a ring-shaped one, or more or less identifiable spots, often weakened but sometimes amplified. In cases where the source quasar presented luminosity "jumps," astronomers were able to measure the different times light took to reach them (after traveling through A or through B) and so obtained a detailed model of the phenomenon. This new technique allowed them to evaluate the quasar's absolute distance in a particularly accurate way. The diagram in figure 15.3 shows an "optimal" configuration for a gravitational mirage to take place. Notice that a gravitational lens is necessarily divergent, since the closer the light beam is to the deflector, the more significant the deflection.

Early on, as soon as he had foreseen the possibility of light

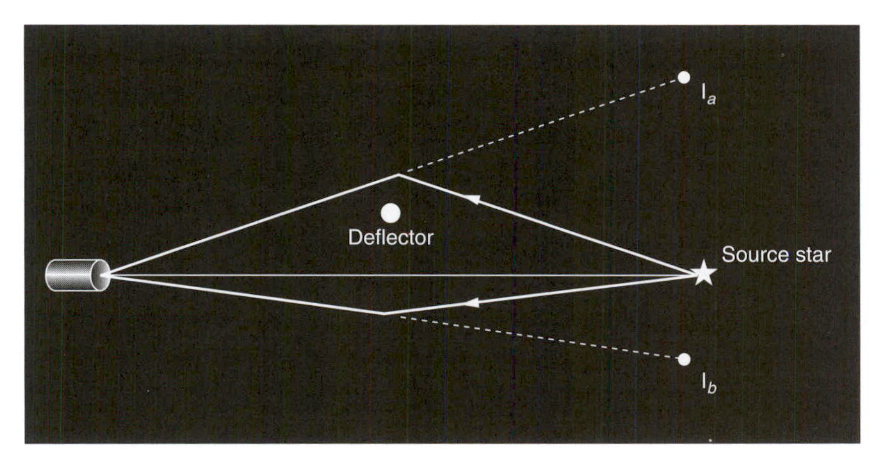

Figure 15.3. Diagram showing a gravitational mirage. The deflector plays the role of a gravitational lens, and the observer on Earth reconstructs the actual light beam (full lines) as though it were a straight line (dotted lines). The diagram shows the source star as well as both shifted images, I_a and I_b.

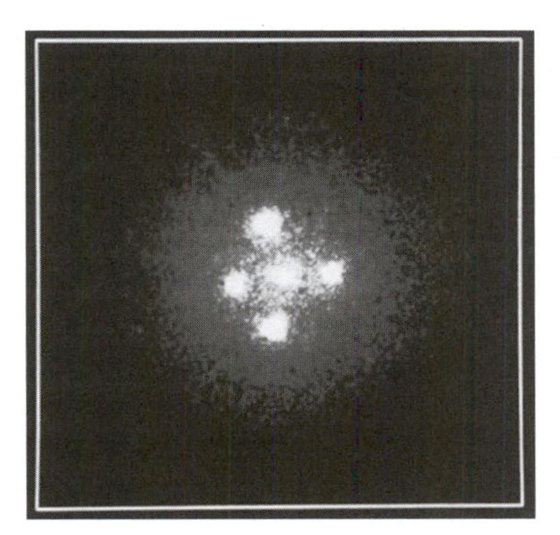

Figure 15.4. Einstein's Cross. The external spots are four images of the same object.

being deflected by gravitation, Einstein envisaged this phenomenon and devoted some quite detailed calculations to it. Today, this effect is used as a tool, as a "gravitational telescope" to study the distribution of matter in the universe. In particular, astrophysicists hope to detect gravitational mirages caused by clusters of "dark matter." They have managed to observe mirages due to gravitational microlenses that indicate the presence of brown dwarfs, small stars that are optically invisible.

COSMOLOGY: A NATURAL PLACE TO "THINK" GENERAL RELATIVITY

I have perpetrated something again as well in gravitation theory, which exposes me a bit to the danger of being committed to a mad house.

A. Einstein[1]

In chapter 4, we saw that in 1912 Einstein used Mach's principle as he conceived his project of a theory of gravitation. We followed him as he struggled to write his gravitation field equations in a curved space. He finally formulated those equations in November 1915, and they worked! They are essentially the same ones we still use today, except for one detail: a cosmological constant that in 1917 he believed had to be added to his equations. What happened? Why did he decide to modify his field equations?

Something was still bothering Einstein: Mach's principle, a central issue for him and one that involved the consistency of the whole theoretical structure. On this point he took a somewhat idealistic stand, and he knew it. What did he want? Since his theory enabled him to define space-time by a set of equations, he would have liked that a universe created by a given distribution of matter would be completely determined by such a distribution. "Completely"—this is not a word to be taken lightly. He wanted that, in the absence of matter, there should be no space-time at all—nothing less.

He was annoyed by the fact that an isolated particle in an otherwise empty space could lead to Schwarzschild's solution, for as the star defines the gravitational field around it, it is also supposed to create the entire space-time, all the way to infinity and its Minkowskian structure. But how can a given mass (which may be assumed to be arbitrarily small—an atom, for instance) create at one go the structure of the whole space-time, from its center to infinity? This was not right. Basically, Einstein could not accept that infinite space could have a structure, an existence, without a material support. It is as if absolute space, that ghost, could still exist—be there before the matter that occupies it! He could not take it. He then wrote to Schwarzschild "jokingly," as he put it (a sign that he was aware of the futility of his assumption), that "If I allow all things to vanish from the world, then, following Newton, the Galilean inertial space remains; following my interpretation, however, *nothing* remains."[2]

Nothing? But what did he mean by that? To the question, what is the space-time created by the void? he certainly did not want the answer to be Minkowski's space. He would have liked that, at most, there should be a shapeless space, a sort of "degenerate" space-time, for he could not accept that space could exist independently of the matter it contains. If the universe is empty, Why should it have a structure? Why should it give origin to a space-time? Why in particular should it have a structure extending all the way to infinity?

More precisely, Einstein would have liked that an isolated body in an otherwise empty space should have no inertia at all; that the inertia of any given body should depend on the masses around it, and in particular on the matter forming the distant stars. In brief, Einstein wanted to solve the problem of the origin of inertia. He wanted a physical explanation to the question of Newton's bucket and Foucault's pendulum (see chapter 4): inertia must be a physical phenomenon like gravitation, which it resembles so much. Inertia must be an interaction between this body here at hand and all other bodies in the universe. And so, he thought, an isolated body in a completely empty space would have no inertia. In short, if all fixed stars could be removed, the surface of Newton's bucket would no longer curve upward nor the sling stretch toward infinity; there would be no centrifugal force, no inertia, and no space any more, only a degenerate space-time. For space-time by itself should not possess any property.

Such questions, such demands, clearly stem from an almost metaphysical ideal, simply because, regardless of the answer we give them, we would never be able to verify on-site the structure of an empty space-time. We would have to empty the universe. . . . This is therefore a question of consistency, a metaphysical question.

That is how Einstein jumped on the cosmological bandwagon. He wanted to write the equations of the universe, take a look at the consequences and, if possible, realize his ideal of physics. He wrote to his friend Besso: "At the moment I'm working quite moderately, so I'm feeling nicely well and am living peacefully along without any discord. In gravitation, I'm now looking for the boundary conditions at infinity; it certainly is interesting to consider to what extent a *finite* world exists, that is, a world of naturally measured finite extension in which all inertia is truly relative."[3]

With the help of a colleague, a mathematician, he tried to formulate Mach's principle in a form compatible with his views. They sought to define a degenerate space-time; degenerate in the sense that the universe would no longer have a spatial dimension, while the temporal dimension would be blocked. But that turned out to be impossible. Einstein did not want to be forced to assume that space is Minkowskian at infinity. That would resemble Newton's absolute space too much, and it would amount to giving up his interpretation of Mach's principle, to which he referred as "the relativity of inertia." There was a way out, one which would allow him to avoid those conditions at infinity: to set those conditions himself, "to consider the universe as a *closed continuum as regards its spatial extension.*" Rather than allow the structure of the universe at infinity to be imposed on him, somewhat gratuitously, he would define this structure through his equations.

He had to find the structure of space-time for the real universe. What were the possible models of universe? Before he could apply his field equations to the universe, he had to decide on the form of the "right-hand member," which defines the matter present in the model and hence in the universe. Only then could he solve those equations, and the solution would yield the structure of space-time, the geometry of the universe.

Obviously, Einstein had to propose a hypothesis as to the distribution of matter in the universe, that is, as to its symmetries. If he assumed that the world had a center and a small density but it was empty at infinity, he would encounter the same difficulties Newton did: the particles forming the universe would leave the system; they would evaporate. Even if the universe was locally somewhat irregular but spatially homogenous on a large scale, he would assume that the density of matter in the universe is, on average, constant,[4] in the same way as "a potato's surface resembles a sphere's surface," he wrote. Actually, he had no choice, because by supposing a more realistic distribution of matter, he would not be able to solve his field equations; it would be too complicated. He needed to assume that the density of matter in the universe was constant, that there are as many stars (on average) here as everywhere. It was a manifestly strange hypothesis, which may even appear unrealistic insofar as the universe is clearly not homogeneous and there is no reason for it to be so. But mathematically it was the simplest hypothesis, one that had been already proposed by William Herschel at the end of the eighteenth century and was an immediately obvious one. In addition, Einstein would not allow this everywhere constant density to change with time; he would assume that the universe was the same, not just everywhere but also forever, that it did not change, that it was static. Finally, he required that his model be closed, which means that its spatial extension was finite, in the same way a sphere, for instance, is finite with respect to three-dimensional space.

But there was a problem, for his imposed conditions were too strong and he soon realized that his fundamental equations had no solution within those constraints. And so, in order to obtain a closed, homogeneous, and static cosmological solution, he was compelled to slightly modify his equations by the introduction of an additional term, λ, which would later be known as "the cosmological constant." For Einstein, this modification of the fundamental equations of his theory was justified by the fact that it allowed him to satisfy his requirements. This is why, when homogeneous solutions were found (and accepted by him) that were also dynamic—that is, they change with time—but without a cosmological constant, he would regret having introduced λ.

This model of the universe in general relativity, or "Einstein universe," was the first cosmological structure ever constructed as a solution to the universe's material contents, and it marked the birth of relativistic cosmology. From then on, the universe was definitely a function of the matter it contains, and it was no longer necessary to set a priori conditions on its structure at infinity, for this structure necessarily followed.

Very soon though, de Sitter would exhibit another model of the universe, a solution of the same problem but without matter: an empty space whose space-time structure was different from that of Minkowski's. This meant that there exist space-times that are at the same time empty and curved, a fact that Einstein refused to accept, for what would then be the origin of their curvature? He attempted to prove, with rather dubious arguments, that de Sitter's space-time created some insurmountable problems. As for Mach's question, the problem of the relativity of inertia, it was not really solved, and Einstein would often return to it without much success. Such is the subject of "Cosmological Considerations on the General Theory of Relativity," in which Einstein invented the universe and opened up an entirely new chapter in the history of the cosmos.

A SPECULATIVE THEORY

The 1920s witnessed a tremendous upheaval in the way the physical universe was viewed. We have seen how Einstein simplified the universe by assuming it to be identical everywhere and forever. Henceforth, astronomers would study its contents, its material structure. What is the shape of the universe? How big is it? Is it essentially composed of our Milky Way, as had been believed to be the case since the eighteenth century, or is our galaxy simply one among many others? It took Edwin Powell Hubble (from Mount Wilson Observatory) to render the existence of other galaxies inevitable. In 1925, he announced the discovery of Cepheid variables in several spiral nebulae; these are variable stars whose distance may be quite accurately determined and therefore also the distance of the nebula to which they belong. The extragalactic nature of nebulae (today's galaxies) was soon beyond doubt. In 1929, Hubble published a first diagram relating galaxies to their speed and establishing the law named after him. It appears that, the more distant the galaxy, the higher its recession speed; but that speed is relative to other galaxies, a speed of the expansion of the universe itself. To visualize this phenomenon, imagine our galaxies as small dots more or less uniformly distributed on the surface of a balloon representing the universe that is being inflated. The relative distance between the galactic dots as well as their relative speed will increase at the same time as the radius of the balloon/universe does; the entire "universe" is expanding. Hence, the universe would not be static, as Einstein had assumed, but dynamic, that is, in expansion. Throughout the 1920s, the work of Aleksander Friedman, the abbot Georges Lemaître, and Howard Robertson on relativistic cosmology had established the theoretical possibility of the existence of dynamic universes. Confronted with their arguments, Einstein had to give in.

The links between cosmology and observation remained extremely tenuous for a long time. Like general relativity, cosmology was a theory ahead of its time. As for the astronomers who should have been the natural users of this new theory of gravitation, they remained impervious to its appeal and sophisticated techniques. They probably knew that general relativity could explain the advance of Mercury's perihelion; perhaps they also knew the formula for the second or even the third test, but that was all. Between 1925 and 1960, no astronomy treatise devoted more than three short pages to the theory. Here is the perception of an eminent relativist, John L. Synge, at the beginning of the 1950s: "But the change of thought is very slow, and it is safe to say that at the present time there are very few astronomers indeed who frame their thoughts about gravitation along the lines of the general theory of relativity. Thus, they 'neglect' it, holding that, within the limits of observational accuracy, its conclusions are indistinguishable from those of the Newtonian theory except in certain critical cases."[5] This was serious, because it indicated that the theory was not recognized, not even at the conceptual level, by its potential users. And it was particularly serious for the relativists, for it meant that the specialists in gravitation were, in practice, those working in celestial mechanics and Newtonian theory, and it was to them that astronomers still turned.

Even if "relativistic cosmology" conceptually and technically depended almost entirely on general relativity, in practice it remained essentially different. It was frequently ignored in the textbooks, and cosmology was often considered a field independent from other aspects of general relativity. On the one hand, the observational factors on which Hubble's law was based were extremely difficult to measure and, on the empirical front, cosmology only added very little, if anything, to general relativity. These facts may explain the limited confidence, if not the limited interest, shown by many relativists in the cosmological field and also the precautions taken by those working in it. To the extremely wide latitude that cosmology was granted by observation, one must add the proximity of "philosophy" casting a—diabolical, for many physicists—shadow on a field considered marginal and suspicious: a speculative theory.

But, paradoxically, it was precisely because of its speculative character that cosmology played an important role in the development of general relativity; for from the beginning, cosmology posed the question of the structure of the universe. Unlike the other domains of general relativity, with its fundamental solutions that could be understood from a neo-Newtonian perspective, cosmology was a field in which it was necessary and possible to "think" general relativity in the context of a structure of space-time *truly* different from the classical one. Einstein's and de Sitter's universes, to mention only the first space-times whose introduction was due to cosmology, really were curved, with a structure completely different from that of Minkowski's. In this respect, relativistic cosmology represented the only consequence of general relativity truly independent from Newton's theory.

In fact, until the beginning of the 1960s, cosmology was the only field where general relativity could be "tested," thought, and pushed to its ultimate consequences in the context of a space-time freed of its Newtonian constraints, that is, in a curved space. It was not by chance that many distinguished cosmologists were behind the overhaul of the interpretation of Schwarzschild's solution, in particular of its "singularity." Cosmology would also make an essential contribution to the recent development of the theory. It was after all easier to imagine the curvature of space than that of a star, to accept a space-time curved on a large scale rather than locally. Is it possible that one of the reasons for the reluctance to interpret Schwarzschild's solution as a black hole came from the difficulty in accepting that the space-time describing the geometry of a star could be curved, twisted; that it was very different from what we observe down here on Earth? Cosmology helped with the understanding and acceptance of those ideas.

THE STANDARD MODEL

Since the 1930s, one cosmological model gradually stood out as the right tool for describing the geometry of the universe: the so-called "standard" model. It is in fact a class of models known as "FRW"—initials of the cosmologists (Friedmann, Robertson, and Walker) who formulated them—which account for everything we know about the observable universe in an accurate way. This class of models corresponds to a homogeneous and isotropic density of matter. A universe of this kind is the same for all observers, regardless of their location; it is dynamic and may be either expanding or contracting. In addition, it possesses a universal time, an uncommon feature in relativity. If we want to understand what happens in an expanding universe, we must start by shrinking the "balloon" that represents the universe until it becomes a single point: the beginning of time, the birth of the universe where space-time is singular. Cosmological matter, even if it changes throughout its history (but we shall not discuss physical cosmology here), is conserved, and therefore all the matter in the universe is supposed to be contained in (in fact, on) our balloon, whose density becomes infinite when its "radius" is zero—at the instant of the big bang.

The fundamental parameter of these models is the "Hubble constant," which represents the speed of expansion of the universe, that is, of the galaxies. The redshift of a galaxy is a measure of its recession speed which, according to the models, is proportional to its distance from us. It is a speed divided by a distance and expressed in kilometers per second per megaparsec.[6] The farther away the galaxy, the higher its recession speed, and the more rapidly it recedes from us: the universe is expanding. Thanks to spectrometric measurements, a galaxy's redshift can be fairly accurately estimated. But calculating its distance is an altogether different story, and measuring Hubble's constant, presently estimated at between 50 and 100 kilometers per second per megaparsec, was not an easy task for our astronomers. Plotting the distance of the galaxies on the x-axis and their speed on the y-axis produces (approximately) a straight line whose slope is the Hubble constant. From this diagram we may deduce the date of the big bang and therefore the "age of the universe." It is believed that the universe is some fifteen billion years old, but this is only an order of magnitude used by astronomers, for example, to compare the age of the universe with our geological timescales. The obvious fact that the Earth should be younger than the universe created some difficult problems not so long ago: since the then current estimate for the Hubble constant was much higher, the age of the universe (the reciprocal of the constant) turned out to be much smaller than the figure generally accepted today.

In 1965, Arno Penzias and Robert Wilson discovered the "background" radiation produced during uncoupling, that is, the moment photons pulled away from the primordial soup. This background radiation exhibited essentially the same intensity and energy spectrum in all directions, and its temperature was that of a black body, that is, 3°K. In fact, the models of the universe then available—in particular thanks to the work of George Gamow—predicted such a phenomenon. What Penzias and Wilson observed was a fossil radiation, a very early image of the universe: the primeval universe. This was probably the most important observation in the history of cosmology after nucleosynthesis and expansion. Until then, only the tally of galaxies, galaxy clusters, and superclusters related to (very uncertain) measurements of their distances supported the assumed homogeneity of the universe, a hypothesis that, surprisingly, became better and better confirmed by observation over the years.

But it was also believed that galaxies were formed by accretion, by gravitational amplification around small nonhomogeneous spots or primordial density fluctuations. On the one hand, the standard model was based on the hypothesis of the large-scale homogeneity of the universe; but, on the other hand, in order to account for the present distribution of matter, it was expected that a lack of homogeneity of the primeval matter, from which the accretion of galaxies would have developed, would be observed.

The COBE (Cosmic Background Explorer) mission's early results made possible an extremely precise measurement of the 3°K radiation (actually, 2.73°K) without detecting any anisotropy. The high homogeneity of the primordial distribution of matter did not allow astrophysicists to propose even the beginning of an explanation as to the formation of galaxies. This situation radically changed when, in April 1992, COBE's team announced the observation of "ripples," or temperature fluctuations of the order of 5×10^{-6} compatible with galaxy formation. The physics of the primordial universe could then really begin. It is one of the questions that dominates contemporary research, in particular, inflation theory.

TOWARD A QUANTUM THEORY OF GRAVITATION

Throughout this book, I have not concealed the essential incompleteness of general relativity: the fact that, even if it was, to the relativists great satisfaction, an excellent theory of gravitation, it did not take into account the quantum world. This was, in a certain sense, a major cause of concern for Einstein who, from the 1920s, had abandoned his theory of gravitation to attempt to construct a unified theory of electromagnetism and gravitation—the quantum domain was not yet on the agenda.

The quantum domain now composes the most significant part of contemporary physics, from a theoretical and technical—if not also political—point of view: not only because it dominates the realm of the infinitely small, but also because it involves a host of industrial techniques and applications that are essential to the contemporary world. From such a perspective, general relativity is extremely isolated. The *only* technical application of which it has ever been a part is the Global Positioning System (GPS),[7] a system of satellites by which the position of an object on Earth may be known with very high precision (one meter, in its military configuration). This involves certain relativistic effects—very weak ones, in actuality, but which nevertheless are regarded today as considerable. Relativists never fail to mention this application of general relativity. Their almost youthful enthusiasm in stressing the single technical application involving, even if only marginally, their theory amounts to an admission of their technical and economic powerlessness and their hope of getting, at last, a piece of the dividends—a hope that existed for a long time and, as we have seen, Synge had already denounced in the foreword of his book at the beginning of the 1960s.

Everything within the province of the infinitely small is beyond general relativity, from electromagnetism to strong and weak interactions; in short, the entire quantum world that "the grand unified theory" describes rather well, even if the "unification" is not perfect. But, conversely, both the grand unified theory and its components ignore everything about gravitation and black holes. This is a fundamental flaw of theoretical physics stemming from the fact that, as we have seen, general relativity is based on a curved non-Euclidean space-time, whereas all quantum theories are based on Minkowski's space-time. The consequences of this difference are extremely serious, for it is much more difficult to work in a curved space than in a pseudo-Euclidean one.

The next step would be a "theory of everything" unifying the quantum and gravitational domains. As regards general relativity, there is an additional reason for seeking this unification: the problems posed by its singularities. Insofar as those singularities involve infinite density, which does not make physical sense, they mark a limit of validity for the theory and the place where quantum phenomena can no longer be ignored, for at this stage they become crucial.

The three fundamental physical constants, G, c, and $\hbar$, must necessarily be at the base of such unification and in particular of a quantum theory of gravitation. Each of these three constants has a specific dimension (length divided by time for the speed of light, for instance), and there is only one combination of them whose dimension is length, the so-called Planck length. Specialists expect this length to be typical of the quantum gravitational field. They also believe that, before Planck length (10^{-33} cm) is reached, none of the classical theories (i.e., general relativity and the grand unification) will be valid. The value of the Planck density at that point, 10^{93} g/cm^3, which is 10^{80} times higher than the nuclear density, shows, as if this would be necessary, the gap—that word is not strong enough—rather, the ocean, the universe, separating the two "physics." Thus, it is at that place that the domain of the future theory would begin—this unified theory of all physical interactions, spanning the universe of the infinitely small (i.e., the weak and strong electromagnetic interactions) and that of the infinitely large structured by gravitation.

Some theoreticians believe that in a quantum theory of gravitation the microscopic structure of space-time would no longer be a four-dimensional continuum, that the very geometry of space-time would be "quantized." Such a unified theory of the gravitational field would be valid at both short and long range and would behave in a certain sense like a serpent biting its tail. . . .

Box 6. Orders of Magnitude

For this calculation of orders of magnitude, let us remain within the Newtonian context and try to get a rough idea of the intensity of the gravitational field when more and more violent processes, from the gravitational standpoint, take place.

Let us calculate the acceleration of gravity on the surface of stars of increasing density. We know that the acceleration of gravity on the Earth's surface, g, is equal to 9.81 ms^{-2}. It is not difficult to calculate this acceleration which, let us recall, is that to which all bodies are subject regardless of their mass, according to the principle of equivalence.

On the surface of a star, the acceleration is none other than $a = -MG/R^2$, where G is of course the (universal) gravitational constant, M the mass of the body creating the gravitational field, and R the star's radius. The acceleration of gravity on the surface of the Sun is of the order of $3g$, that is, three times that on the Earth. On a white dwarf (whose typical mass is equivalent to the Sun's), this acceleration is ten thousand times stronger than on Earth. This is due to the fact that the radius of a white dwarf is about one hundred times smaller than the Sun's. As for the value of this acceleration on a neutron star, it is (roughly) $10^{10}g$, or ten billion times stronger than on Earth, computed in Newtonian terms, which is not too far off in this case. These results are not surprising because, typically, the mass of a neutron star or a white dwarf is close to that of the Sun whereas its radius is quite different. And since, according to Newton's law, the radius appears squared in the denominator of the acceleration formula, the acceleration increases "geometrically": it is proportional to $1/R^2$.

It does not make much sense to calculate, from a Newtonian point of view, the acceleration of gravity of a solar mass on the surface of a black hole. But we may imagine that this acceleration must be greater by a factor of 10, given that the radius of a black hole is three times smaller than that of a neutron star of comparable mass.

(*continued*)

Box 6. *(continued)*

Conversely, if we calculate the acceleration of gravity on a super giant star such as Bételgeuse, we would be disappointed: it is one thousand times weaker than the acceleration of gravity on Earth. Obviously, its density is very small, a million times smaller than that of the Sun. On the contrary, the density of a white dwarf is one million times that of the Sun, and the density of a neutron star is greater than the Sun's by a factor of 10^{15} (one million billion times more dense). This is due to the fact that the acceleration is directly proportional to the density of the star.

NOTES

1. Einstein to P. Ehrenfest, 4 February 1917, German original in *CPAE*, vol. 8, p. 386; English translation, vol. 8, p. 282.

2. Einstein to K. Schwarzschild, 9 January 1916, German original in *CPAE*, vol. 8, p. 241; English translation, vol. 8, p. 176.

3. Einstein to M. Besso, 14 May 1916, German original in *CPAE*, vol. 8, p. 287.

4. At the time, Einstein was only interested in spatial variations of density; he could not imagine that the universe may change with time and did not allow it. That is fundamentally the reason why he was so reluctant to accept Friedmann's solution. In this regard, cf. Eisenstaedt, 1993, for a historical analysis.

5. Synge, 1952, p. 304.

6. A megaparsec (Mpc) is an astronomical unit of measure equivalent to 3.09×10^{22} meters.

7. See Hofmann-Wellenhof et al., 1994.

AFTERWORD

The Paths of General Relativity

ACCORDING TO THE publisher of the French dictionary *Robert*, there are two sorts of languages: those that are dead and those that are sick. Latin is considered a dead language, for it has ceased evolving; it does not integrate any new elements nor does it reject old sediments or foreign words, and because it is dead it cannot be sick.

One consequence of this somewhat provocative idea is that dead people are the only ones that are not sick, just as, according to the famous paradox, only a clock that has stopped working tells the right time at least once a day. And so, as we know only too well on this side of the Channel, French has been "sickened" by English, which has been "sickened" by American English, which itself is about to get "sick."

An honest physicist would adapt this paradox by claiming that there are two kinds of theories: the dynamic theories, those that are alive and therefore necessarily afflicted by some ailment, and the static ones, those theories that are dead and nobody pays any attention to anymore, except historians. Unfortunately for the relativists, this is not the case for Newton's theory—and may not be any time soon. It is far from being discarded because it is routinely employed in connection with satellites and in advanced research, such as that regarding the evolution and stability of the solar system. From that point of view, Newtonian theory may be considered as an excellent approximation to its relativist sister, and so more than adequate for the everyday needs of numerous astronomers and other specialists in celestial mechanics.

That brings us back to Einstein, who believed that every theory, and in particular general relativity, would cease being alive the moment it was replaced by a more precise and general one.

General relativity was for Einstein an interesting but unsatisfactory theory: a pause along the path toward new theoretical horizons, horizons that burgeoned profusely but had not yet blossomed.

Is general relativity sick? Of course it is sick, because it is still growing and necessarily suffering from many illnesses, in particular those typical of adolescence. There are plenty of doctors at its bedside. Some develop remedies to cure it of its minor ailments—the questions that remain open, such as those regarding the singularities, for instance. Others, more daring, are working on metatheories that they hope will surpass it; they dream of killing it so that their theory can take its place, a theory unifying the quantum and gravitational domains and that at the same time would cure physics of its incompleteness and divisions. It is the dream of every ambitious theoretician.

Let us ignore the pejorative character of the term *sickness* and interpret it rather as a sign of life, of a struggle for life, as a proof that Einstein's theory is not dead. Has it ever been closer to death than during the years of wandering in the desert that I discussed in chapter 10? The theory was then paralyzed, bogged down by three tests that appeared to represent also the end of it. Certain experts believed, hoped, that it was already a "historical" theory, dead before having really lived. That view was a serious mistake! General relativity was only suffering from growing pains from which it recovered nicely.

General relativity is in fact at the origin of an extraordinarily dynamic field of physics: high-energy gravitation, which involves pulsars, quasars, black holes, gravitational waves, the big bang and, of course, it is also at the origin of cosmology. That is its major contribution to contemporary physics, and it has also made its mark in the rapidly growing field of "astroparticles."

All these phenomena keep bringing forth new challenges that general relativity will not be able to take on forever. It will die—I'm jumping ahead here—like a supernova, in a kind of high-powered fireworks, after being confronted by some quite extraordinary observations. The theories that have been important for us do not go away quietly, they die following a period of intense activity, like supernovae. A day will come that will see the end of general relativity, but it will be a spectacular end, brought about by some experiment that will not be "crucial" in vain, because it will be mark the arrival of a better theory. Every healthy theory lives on borrowed time; after all, theories are the work of humans.

But for now, the theory lives on. After a glorious childhood, an ungrateful adolescence, and a striking renewal, general relativity has now reached middle age. It has made a comeback on the international stage of physics and the best theoreticians are experts in it. A growing number of increasingly precise observations—and experiments—enrich its domain. For the first time in its life, its verification has produced a Nobel Prize. From its geometric pedestal—and to everyone's surprise—it proudly, if not without a bit of insolence and irony, dominates the observational field as never before. It is the subject of highly specialized research that contributes to the renewal of theoretical physics. To its detractors of yesterday, it appears to say: "You see, I needed time to grow up. . . . It is now your turn do some thinking!" It is presently at the heart of the theoretical field from which a unified supertheory of the quantum and gravitational domains should emerge.

In short, all is well in relativity's kingdom. But we have to get used to the idea. . . . Much work remains to be done!

Bibliography

Bergmann P. G. 1942. *Introduction to the Theory of Relativity*. Foreword by A. Einstein. New York: Prentice-Hall. Reprint, New York: Dover, 1976.

Boi, L. 1995. *Le Problème mathématique de l'Espace*. Berlin: Springer.

Bondi, H. 1962. "A discussion on the present state of relativity." *Proceedings of the Royal Society of London A*, 270, pp. 297–356.

Born, M. 1956. "Physics and Relativity." In Mercier and Kervaire, 1956, pp. 244–260.

———. ed. 1969. *Albert Einstein–Hedwig und Max Born. Briefwechsel 1916–1955*. Munich: Nymphenburger. Translated by Irene Born as *The Born-Einstein Letters. Correspondence between Albert Einstein and Max and Hedwig Born from 1916 to 1955. With commentaries by Max Born*. London: Macmillan, 1971.

Bosscha, J., ed. 1900. Collection of papers offered by the authors to H. A. Lorentz, Professor of Physics at the University of Leiden, on the 25th anniversary of his PhD, 11 December 1900. *Archives néerlandaises des sciences exactes et naturelles*, 5.

Bouasse, H. 1923. *La question préalable contre la théorie d'Einstein*. Paris: Albert Blanchard.

Bouguer, P. 1749. *La Figure de la Terre, déterminée par les observations de Messieurs Bouguer & de la Condamine, de l'Académie royale des sciences, envoyés par ordre du Roy au Pérou, pour observer aux environs de l'équateur*. Paris: Jombert.

Bourdieu, P. 1976. "Le champ scientifique." *Actes de la Recherche en Sciences Sociales*, 2/3, pp. 88–104.

Brian, D. 1996. *Einstein: A Life*. New York: John Wiley and Sons.

Buchwald, J. Z. 1989. *The Rise of the Wave Theory of Light*. Chicago: The University of Chicago Press.

Casimir, H.B.G. 1983. *Haphazard Reality: Half a Century of Science*. New York: Harper and Row.

Cassini de Thury, C.-F. 1744. *La Méridienne de l'Observatoire de Paris, vérifiée dans toute l'étendue du Royaume par de nouvelles opérations*. Paris: H. L. Guérin & J. Guérin.

Caveing, M., and B. Vitrac, eds. 1990. *Les Eléments d'Euclide d'Alexandrie, traduits du texte de Heiberg*. Paris: Presses Universitaires de France.

Chandrasekhar, S. 1969. "The Richtmyer memorial lecture—some historical notes." *American Journal of Physics*, 37, pp. 577–584.

———. 1979. "Einstein and general relativity: historical perspectives." *American Journal of Physics*, 47, pp. 212–217.

Christianson, G. E. 1984. *In the Presence of the Creator: Isaac Newton and His Times*. New York: Free Press.

Connes, A., A. Lichnerowicz, and M. P. Schützenberger. 2000. *Triangle de pensées*. Paris: Jacob.

Corry, L., J. Renn, and J. Stachel. 1997. "Belated decision in the Hilbert-Einstein priority dispute." *Science*, 278, pp. 1270–1273.

CPAE. The Collected Papers of Albert Einstein. Edited by J. Stachel et al. Translated by Anna Beck. 8 vols. Princeton, NJ: Princeton University Press, 1987–.

Crelinsten, J. 1980. "Physicists receive relativity: Revolution and reaction." *The Physics Teacher*, 18, pp. 187–193.

———. 1981. "The Reception of General Relativity among American Astronomers: 1910–1930." PhD diss. University of Montreal.

———. 1984. "William Wallace Campbell and the 'Einstein problem': An observational astronomer confronts the theory of relativity." *Historical Studies in the Physical Sciences*, 14, pp. 1–91.

Crommelin, A.C.D. 1919a. "The eclipse expedition to Sobral." *The Observatory*, 42, pp. 368–371.

———. 1919b. "The eclipse of the sun on May 29." *Nature*, 102, pp. 444–446.

Curtis, H. D. 1917. "Space, time and gravitation." *Astronomical Society of the Pacific*, 29, pp. 63–64.

Darrigol, O. 1995. "Henri Poincaré's criticism of fin de siècle electrodynamics." *Studies in History and Philosophy of Modern Physics*, 26, pp.1–44.

De Witt, C., and B. De Witt, eds. 1973. *Black Holes*. New York: Gordon and Breach.

Dicke, R. H. 1957. "Gravitation without a principle of equivalence." *Reviews of Modern Physics*, 29, pp. 363–376.

———. 1962. "Mach's principle and equivalence." In *Evidence for Gravitational Theories. Proceedings of the International School of Physics "Enrico Fermi,"* course 20 Varenna, 1961, ed. C. Møller. New York: Academic Press.

———. 1964. *The Theoretical Significance of Experimental Relativity.* New York: Gordon and Breach.

Dreyer, John L. E. 1912. Introduction to *The Scientific Papers of Sir William Herschel.* 2 vols. London: The Royal Society and the Royal Astronomical Society.

Dyson, F. W., A. S. Eddington, and C. Davidson. 1920. "A determination of the deflection of light by the Sun's gravitational field from observations made at the total eclipse of May 29, 1919." *Philosophical Transactions of the Royal Society of London A,* 220, pp. 291–333.

Earman J., and C. Glymour. 1980a. "Relativity and eclipses: the British eclipse expeditions of 1919 and their predecessors." *Historical Studies in the Physical Sciences,* 11, pp. 49–85.

———. 1980b. "The gravitational red shift as a test of general relativity: history and analysis." *Studies in History and Philosophy of Science,* 11, pp. 175–214.

Eddington A. S. 1926. *The Internal Constitution of the Stars.* Cambridge: Cambridge University Press.

Einstein, A. 1905a. "Zur Elektrodynamik bewegter Körper." *Annalen der Physik* 17: 891–921. English translation in Lorentz et al., 1923, pp. 35–65.

———. 1905b. "Ist die Trägheit eines Körpers von seinem Energieinhalt abhängig?" *Annalen der Physik,* 18, pp. 639–641. English translation in Lorentz et al., 1923, pp. 67–71.

———. 1911. "Über den Einfluss der Schwerkraft auf die Ausbreitung des Lichtes." *Annalen der Physik,* 35, pp. 898–908. English translation in Einstein et al., 1923, pp. 97–108.

———. 1915. "Zur allgemeinen Relativitätstheorie." *Sitzungsberichte der Königlich Preussichen Akademie der Wissenschaften zu Berlin*: pp. 778–786.

———. 1916. "Die Grundlage der allgemeinen Relativitätstheorie." *Annalen der Physik,* 49, pp. 769–822. English translation in Einstein et al., 1923, pp. 109–164.

———. 1919a. "Grundgedanken und Methoden der Relativitäts-Theorie in ihrer Entwicklung dargestellt." Manuscript 2-070-23. The Pierpont Morgan Library. New York, NY.

———. 1919b. "Induktion und Deduktion in der Physik." *Berliner Tageblatt*, 25 December, Suppl. 4.

———. 1921a. "A brief outline of the development of the theory of relativity." *Nature*, 106, pp. 782–784.

———. 1921b. "Geometrie und Erfahrung." *Sitzungsberichte der Preussischen Akademie der Wissenschaften zu Berlin*: pp. 123–130.

———. 1922. "Theoretische Bemerkungen zur Supraleitung der Metalle." In *Het Natuurkundig Laboratorium der Rijksuniversiteit te Leiden in de Jaren 1904–1922. Gedenkenboek aangeboden aan H. Kammerlingh Onnes*, 11 March 1922, Leiden: E. Ijdo. pp. 429–435.

———. 1929. "The new field theory." *Times* (London), 4 February 1929.

———. 1930. "Théorie unitaire du champ physique." *Institut Poincaré Annales*, 1, pp. 1–24.

———. 1949a. "Autobiographical notes." In Schilpp ed. 1949, vol. 1, pp. 2–94.

———. 1949b. "Reply to criticisms." In Schilpp 1949, vol. 2, pp. 663–693.

———. 1950. "On the generalized theory of gravitation." *Scientific American*, 188, pp. 13–17.

———. 1954. *Ideas and Opinions*. Translated by Sonja Bargmann, New York: Crown, 1982.

———. 1956. *Lettres à Maurice Solovine*. Paris: Gauthier-Villars.

———. 1989–1993, *Oeuvres choisies*. Edited by F. Balibar. 6 vols. Paris: Éditions du Seuil.

Eisenstaedt, J. 1982. "Histoire et singularités de la solution de Schwarzschild (1915–1923)." *Archive for History of Exact Sciences*, 27, pp. 157–198.

———. 1986. "La relativité générale à l'étiage, 1925–1955." *Archive for History of Exact Sciences*, 35, pp. 115–185.

———. 1987. "Trajectoires et impasses de la solution de Schwarzschild." *Archive for History of Exact Sciences*, 37, pp. 275–357.

———. 1991. "De l'influence de la gravitation sur la propagation de la lumière en théorie newtonienne. L'archéologie des trous noirs." *Archive for History of Exact Sciences*, 42, pp. 315–386.

———. 1993. "Cosmologie." In *Albert Einstein: Oeuvres choisies*, ed. F. Balibar, vol. 3, pp. 83–129. Paris: Seuil.

———. 1997. "Laplace: l'ambition unitaire ou les lumières de l'astronomie." *Académie des Sciences Paris Comptes Rendus Série II*, 324, pp. 565–574.

Eötvös, R. von. 1874. "Über die Anziehung der Erde auf verschiedene Substanzen." *Mathematische und naturwissenschaftliche Berichte aus Ungarn*, 8, pp. 65–68.

Feuer, L. S. 1982. *Einstein and the Generations of Science*. 2d ed. p. 87. New York: Basic Books.

Feynman, R. P. et al. 1995. *Feynman Lectures on Gravitation*. Edited by B. Hatfield. Foreword by J. Preskill and K. S. Thorne. Reading, MA: Addison-Wesley.

Fölsing, A. 1997. *Albert Einstein: A Biography*. New York: Viking.

Foucault, L. 1851. "Démonstrations physiques du mouvement de rotation de la Terre au moyen du pendule." *Académie des Sciences (Paris). Comptes Rendus*, 32, pp. 135–138.

Fouchy, G. de. 1751. "Sur l'aberration de la lumière des planètes et des comètes." *Académie Royale des Sciences (Paris), Histoire pour 1746*, pp. 101–104.

Frank, P. 1950. *Einstein sa vie et son temps*. Paris: Albin Michel.

Gillmor, C. S. 1971. *Coulomb and the Evolution of Physics and Engineering in Eighteenth-Century France*. Princeton, N.J.: Princeton University Press.

Goenner, H. 1984. "Unified field theories: From Eddington and Einstein up to now." In *Proceedings of the International Conference on Relativity and Cosmology*, ed. V. DeSabbata, and T. M. Karade. Singapore: World Scientific.

Gold, T. 1965. "Summary of after-dinner speech." In *Quasi-Stellar Sources and Gravitational Collapse, Proceedings of the First Texas Symposium of Relativistic Astrophysics*, 1964, ed. Robinson I., A. Schild, and E. L. Schücking. Chicago: University of Chicago Press.

Hakim, R. 1994. *Gravitation relativiste*. Paris: CNRS Editions.

Harrison, B. K., K. S. Thorne, M. Wakano, and J. A. Wheeler. 1965. *Gravitation Theory and Gravitational Collapse*. Chicago: University of Chicago Press.

Harrison, Edward R. 1987. *Darkness at Night*. Cambridge, MA: Harvard University Press.

Hawking, S. W. and G.F.R. Ellis. 1973. *The Large Scale Structure of Space-Time*. Cambridge: Cambridge University Press.

Heath, Thomas L., ed. 1956. *The Thirteen Books of Euclid's Elements*. Translated by Sir Thomas L. Heath. New York: Dover.

Hentschel, K. 1994. "Erwin Finlay Freundlich and testing Einstein's theory of relativity." *Archive for History of Exact Sciences*, 47, pp. 143–201.

Hofmann-Wellenhof, B., H. Lichtenegger, and J. Collins. 1994. *Global Positioning System*. 3d ed. New York: Springer-Verlag.

Holton, G. 1973. *Thematic Origins of Scientific Thought: Kepler to Einstein*. Cambridge, MA: Harvard University Press.

———. 1981. *L'Imagination scientifique*. Translated into French by J. F. Roberts. Paris: Gallimard.

Holton, G., and Y. Elkana, eds. 1982. *Albert Einstein, Historical and Cultural Perspectives: The Centennial Symposium in Jerusalem*. Princeton: Princeton University Press.

Howard, D., and J. Stachel, eds. 1989. *Einstein and the History of General Relativity*. Proceedings of the 1986 Osgood Hill Conference. Einstein Studies, vol. 1. Boston: Birkhauser.

Hoyle, C. D. et al. 2001. "Sub-millimeter tests of the gravitational inverse-square law: A search for 'large' extra dimensions." *Physical Review Letters*, 86, pp. 1418–1421.

Infeld, L. 1950. *Albert Einstein: His Work and Its Influence in Physics*. New York: Charles Scribner's Sons.

———. ed. 1964. *Conférence internationale sur les théories relativistes de la gravitation*, Warsaw and Jablonna, 25–31 July 1962, Warsaw. Paris: Gauthier-Villars.

Israel, W. 2000. "Black Hole 2000: the Astrophysical Era." *Astronomical Society of the Pacific, Publications*, 112, pp. 583–585.

Jaki, S. L. 1978. "J. G. von Soldner and the gravitational bending of light with an English translation of his essay on it published in 1801." *Foundations of Physics*, 8, pp. 927–950.

Jeans, J. H. 1917. "Einstein's Theory of Gravitation." *The Observatory*, 40, pp. 57–58.

Jungnickel, C., and R. McCormmach. 1999. *Cavendish: The Experimental Life*. Cranbury, NJ: Bucknell.

Katzir, S. 1996. "Gravitation in Poincaré's relativity theory." Preprint.

Kuhn, T. S. 1962. *The Structure of Scientific Revolutions*. Chicago: University of Chicago Press.

Lalande, A. 1960. *Vocabulaire technique et critique de la philosophie*. Paris: Presses Universitaires de France.

Lanczos, C. 1932. "Stellung der Relativitätstheorie zu anderen physikalischen Theorien." *Die Naturwissenschaften*, 20, pp. 113–116.

———. 1955. "Albert Einstein and the theory of relativity." *Il Nuovo Cimento*, 10 (2), Suppl., pp. 1193–1220.

Langevin, P. 1932. "L'oeuvre d'Einstein et l'astronomie." *Astronomie* 45, pp. 277–297.

Laplace, P.-S. 1796. *Exposition du système du monde*. 2 vol. Paris: Imprimerie du Cercle-Social.

———. 1799. "Beweis des Satzes, dass die anziehende Kraft bey einem Weltkörper so gross seyn könne, dass das Licht davon nicht ausströmen kann." *Allgemeine Geographische Ephemeriden* 4, 1–6.

Laue, Max von. 1924. Die Relativitätstheorie. Vol. 2. Die allgemeine Relativitätstheorie. 4th ed. Braunschweig: Friedrich Vieweg und Sohn.

Lévy-Leblond, J. M. 1976. "What is so 'special' about relativity?" In *Group Theoretical Methods in Physics*: Fourth International Colloquium on Group Theoretical Methods in Physics, Nijmegen, Netherlands, 1975. Lecture Notes in Physics, vol. 50, ed. A. Janner, T. Janssen, and M. Boon, pp. 617–627. Berlin: Springer-Verlag.

———. 1977. *Les relativités*. Les Cahiers de Fontenay, vol. 8.

Lorentz, H. A., A. Einstein, H. Minkowki, and H. Weyl. 1923. *The Principle of Relativity: A Collection of Original Memoirs on the Special and General Theory of Relativity*. Notes by A. Sommerfeld. Translated by W. Perrett and G. B. Jeffery. London: Methuen. Reprint, New York: Dover, 1952.

Luminet, J. P. 1987. *Les trous noirs*. Paris: Belfond.

Mach, Ernst. 1883. *Die Mechanik in ihrer Entwickelung: Historisch-kritisch dargestellt*. Leipzig: Brockhaus. Translated by T. J. MacCormack as *The Science of Mechanics: A Critical and Historical Account of Its Development*, 6th ed. La Salle, IL: Open Court, 1960.

———. 1904. *La mécanique*. Translated into French by Émile Bertand. Paris: Hermann. Reprint Sceaux: Gabay, 1987.

McCormmach, R. 1968. "John Michell and Henry Cavendish: Weighing the stars." *The British Journal for the History of Science*, 4, pp. 126–155.

McVittie, G. C. 1956. *General Relativity and Cosmology*. London: Chapman and Hall.

Mercier, A., and M. Kervaire, eds. 1956. *Cinquantenaire de la théorie de la relativité*. Actes du congrès de Berne, 11–16 Juli 1955. *Helvetia Physica Acta, Suppl*. 4.

Michell, J. 1784. "On the Means of Discovering the Distance, Magnitude, &c. of the Fixed Stars, in consequence of the Diminution of the Velocity of their Light, in case such a Diminution should be found to take place in any of them, and such other Data should be procured from Observations, as would be farther necessary for that Purpose." By the Rev. John Michell, B.D.F.R.S. In a letter to Henry Cavendish, Esq. F.R.S. and A.S., *Philosophical Transactions of the Royal Society of London*, 74, pp. 35–57.

Miller, A. I. 1981. *Albert Einstein's Special Theory of Relativity, Emergence (1905) and Early Interpretation (1905–1911)*. Reading, MA: Addison-Wesley.

Milne, E. A. 1935. *Relativity, Gravitation and World-Structure*. Oxford: Clarendon Press.

———. 1940. "Kinematical relativity." *Journal of the London Mathematical Society*, 15, pp. 44–80.

Misner, C. W., K. S. Thorne, and J. A. Wheeler. 1973. *Gravitation*. New York: Freeman.

Newcomb, S. 1882. "Discussion and results of observations on transits of Mecury from 1667 to 1881." *Astronomical Papers Prepared for the Use of the American Ephemeris and Nautical Almanac*, 1, pp. 367–487.

Newton, I., 1687. *Sir Isaac Newton's Mathematical Principles of Natural Philosophy and His System of the World*. Vol. 2: *The System of the World*. Translated by A. Motte and F. Cajori. Berkeley: University of California Press, 1962.

———. 1729. *Sir Isaac Newton's Mathematical Principles of Natural Philosophy and His System of the World*. Translated by A. Motte. Revised by F. Cajori. Reprint, Berkeley: University of California Press, 1962.

———. 1730. *Opticks*. 4th ed. London. Reprint, New York: Dover, 1952.

Nordmann, C. 1922. "Einstein expose et discute sa théorie." *La Revue des Deux Mondes*, 9, pp. 129–166.

Pais, A. 1982. "Subtle is the Lord . . .": *The Science and Life of Albert Einstein*. Oxford: Oxford University Press.

Pauli, W. 1958. *Theory of Relativity*. Translated by G. Field. Oxford: Pergamon Press. Reprint, New York: Dover, 1981.

Penrose, R. 1969. "Gravitational collapse: the role of general relativity." *Rivista del Nuovo Cimento, numero speciale*, 1, pp. 252–276.

Poincaré, H. 1891. "Les géométries non-euclidiennes." *Revue générale des Sciences pures et appliquées*, 2, pp. 769–774. Reprinted with some changes in *La Science et l'Hypothèse*, 1902, pp. 49–67.

———. 1898. "La mesure du temps." *Revue de Métaphysique et Morale*, 6, pp. 1–13.

———. 1902. *La science et l'hypothèse*. Paris: Flammarion.

———. 1905. "Sur la dynamique de l'électron." *Académie des Sciences (Paris). Comptes Rendus*, 140, 1, pp. 1504–1508.

———. 1906. "Sur la dynamique de l'électron." *Circolo Matematico di Palermo. Rendiconti*, 21, pp. 129–176, 494–550.

———. 1953. "Les limites de la loi de Newton." *Bulletin Astronomique (Paris)*, 17, pp. 121–178, 181–269.

Popper, K. 1935. *Logik der Forschung zur Erkenntnistheorie der modernen Wissenschaft*. Wien: Springer. Translation: *The Logic of Scientific Discovery*. London: Routledge, 2002.

Postel-Vinay, O. 2000. "Alain Connes: la réalité mathématique archaïque." *La Recherche*, 332, pp. 109–111.

Ricci, G., and T. Levi-Civita. 1901. "Méthodes de calcul différentiel absolu et leurs applications." *Mathematische Annalen*, 54, pp. 125–201.

Römer, O. C., 1676. "Demonstration touchant le mouvement de la lumière trouvé par M. Römer de l'Academie Royale des Sciences. *Journal des Savants*, 7 décembre 1676, pp. 233–236. Reproduced by S. Débarbat in R. Taton 1978, pp. 151–154. Translation: "A Demonstration concerning the motion of light, communicated from Paris, in the Journal des Sçavants, and here made English." *Philosophical Transactions of the Royal Society of London*, (1677), 12, pp. 893–894.

Rosenfeld, B. A. 1988. *A History of Non-Euclidean Geometry*. Berlin: Springer-Verlag.

Roseveare, N. T. 1982. *Mercury's Perihelion from Le Verrier to Einstein*. Oxford: Clarendon Press.

Royal Astronomical Society. 1919. "Meeting of the Royal Astronomical Society, 1919 July 11." *The Observatory*, 42, pp. 297–299.

Royal Astronomical Society. 1932. "Meeting of the Royal Astronomical Society, 1931 Dec. 11." *The Observatory*, 55, pp. 1–10.

Royal Society and the Royal Astronomical Society. 1919. "Joint Eclipse Meeting of the Royal Society and the Royal Astronomical Society, 1919 November 6." *The Observatory*, 42, pp. 389–398.

Schild, A. 1960. "Equivalence principle and red-shift measurements." *American Journal of Physics*, 28, pp. 778–780.

Schilpp, P. A., ed. 1949. *Albert Einstein: Philosopher-Scientist*. 3d ed., 2 vol. Evanston: IL: The Library of Living Philosophers, 1982.

Schwarzschild, K. 1900. "Über das zulässige Krümmungsmass des Raumes." *Vierteljahrsschrift der Astronomischen Gesellschaft*, 35, pp. 337–347.

See, T.J.J. 1916. "Einstein's theory of gravitation." *The Observatory*, 39, pp. 511–512.

Soldner, Johann G. von. 1801. "Ueber die Ablenkung eines Lichtstrals von seiner geradlinigen Bewegung, durch die Attraktion eines Weltkörpers, an welchem er nahe vorbei geht." *Astronomisches Jahrbuch für das Jahr 1804*, pp. 161–172. English translation in Jaki 1978.

Sommerfeld, A. 1949. "To Einstein's Seventieth Birthday." In vol. 1, Schilpp, 1949, pp. 99–105.

Speziali, P., ed. 1972. *Albert Einstein–Michele Besso. Correspondance 1903–1955*. Paris: Hermann.

Stachel, J. 1974. "The rise and fall of geometrodynamics." In *PSA 1972, Proceedings of the 1972 Biennial Meeting, Philosophy of Science Association*, Boston Studies in the Philosophy of Science, vol. 20, ed. K. F. Schaffner and R. S. Cohen, pp. 31–54. Dordrecht: D. Reidel.

———. 1980. "Einstein and the rigidly rotating disk." In *General Relativity and Gravitation: One Hundred Years After the Birth of Albert Einstein*, ed. A. Held, vol. 1, pp. 1–15. New York: Plenum.

———. 1995. "History of relativity." In: *Twentieth Century Physics*, ed. L. M. Brown, A. Pais, and B. Pippard, vol. 1, pp. 249–356. New York: Institute of Physics.

Synge, J. L. 1952. "Orbits and rays in the gravitational field of a finite sphere according to the theory of A. N. Whitehead." *Proceedings of the Royal Society of London A*, 211, pp. 303–319.

———. 1960. *Relativity: The General Theory*. Amsterdam: North-Holland.

Taton, R. 1957–1964. *Histoire générale des sciences*. 4 vol. Paris: Presses Universitaires de France.

———. 1978. *Roemer et la vitesse de la lumière*. Paris: Vrin.

Thomson, J. J. 1920. "Poynting's scientific papers." *Nature*, 106, pp. 559–561.

Trautman, A. 1966. "The general theory of relativity." *Soviet Physics Uspekhi*, 9, pp. 319–335.

Trumpler, R. J. 1923. "Historical note on the problem of light deflection in the Sun's gravitational field." *Science*, 58, pp. 161–163.

———. 1956. "Observational results on the light deflection and on red-shift in star spectra." In Mercier and Kervaire 1956, pp. 106–113.

Walter, S. 1999. "Minkowski, mathematicians, and the mathematical theory of relativity." In *The Expanding Worlds of General Relativity*, Einstein Studies, vol. 7, ed. H. Goenner, J. Renn, J. Ritter, and T. Sauer, pp. 45–86. Boston: Birkhäuser.

Weyl, H. 1918. *Raum-Zeit-Materie*. Berlin: Julius Springer. *Space-Time-Matter*. Translated from 4th ed., 1921, by H. L. Brose. London: Methuen, 1922. Reprint, New York: Dover, 1950.

Wheeler, J. A. 1957. "The present position of classical relativity theory and some of its problems." In *Conference on the Role of Gravitation in Physics at the University of North Carolina, Chapel Hill*, ed. C. De Witt. WADC Technical Report, 57-216. Wright-Patterson AFB, OH: Wright Air Development Center, U.S. Air Force.

———. 1968. "Our Universe: the Known and the Unknown." *American Scientist*, 56, 1–20.

Whitehead, A. N. 1922. *The Principle of Relativity with Applications to Physical Science*. Cambridge: Cambridge University Press.

Whitrow, G. J. 1967. *Einstein: The Man and His Achievement*. London: British Broadcasting Corporation. Reprint, New York: Dover, 1973.

Whittaker, E. T. 1953. *A History of the Theories of Aether and Electricity. The Modern Theories, 1900–1926*. 2 vols. London: Thomas Nelson and Sons.

Wigner, E. P. 1960. "The unreasonable effectiveness of mathematics in the natural sciences." *Communications on Pure and Applied Mathematics*, 13, pp. 1–14.

Williams, J. G., D. H. Boggs, J. O. Dickey, and W. M. Folkner. 2002. "Lunar Laser tests of gravitational physics." In *The Ninth Marcel Grossmann Meeting on Recent Developments in Theoretical and Experimental General Relativity, Gravitation, and Relativistic Field Theories*, Rome, 2–8 July 2000, ed. V. G. Gurzadyan, R. T. Jantzen, and R. Ruffini, vol. 3, p. 1797. Singapore: World Scientific.

Name Index